PASCAL'S FIRE

PASCAL'S FIRE

SCIENTIFIC FAITH AND RELIGIOUS UNDERSTANDING

KEITH WARD

ONEWORLD

OXFORD

PASCAL'S FIRE

Oneworld Publications
(Sales and Editorial)
185 Banbury Road
Oxford OX2 7AR
England
www.oneworld-publications.com

ISBN-10: 1–85168–446–8
ISBN-13: 978–1–85168–446–5

Cover design by Liz Powell
Typeset by Jayvee, Trivandrum, India
Printed and bound in Great Britain by
Biddles Ltd., King's Lynn

Learn more about Oneworld. Join our mailing list to
find out about our latest titles and special offers at:

www.oneworld-publications.com/newsletter.htm

To Arthur Peacocke

This book is based on a series of lectures given at Gresham College, London, in 2004–5.

23 November 1654: From about half past ten in the evening until half past midnight. Fire. The God of Abraham, the God of Isaac, the God of Jacob, and not of the philosophers and men of science.

CONTENTS

Contents

ACKNOWLEDGEMENTS

I would like to thank my publisher, Novin, and his readers for the great patience with which they have helped this book to come to its final form. I am also greatly indebted to the many patient scientists who have discussed with me issues treated in this book, and especially Arthur Peacocke, John Polkinghorne, Russell Stannard, Paul Davies and John Barrow, who have discussed parts of it in detail at various times. I add the necessary proviso that I do not hold them responsible for my views, but only for inspiring those views, whether they like it or not.

PREFACE

There are many excellent books by scientists on the relationship between science and religion. This is, I fear, not one of them. I am basically a philosopher, who later turned, by accident, into a theologian. Like most classical philosophers, I believe there is a God, the creator of the universe. This book argues that if you accept all the well-established findings of modern science, then you will naturally start asking questions about God. Moreover, the way we think about God today has to be seriously affected by close attention to the findings of modern science. Not all scientists believe in God, as any of the beautifully written books by Richard Dawkins, Britain's best-known evangelical atheist, will show. But modern science does lead to the question of God (as Dawkins's books also show). And I aim to show that some of the things it suggests about God are, surprisingly to some people, very positive and illuminating.

Nevertheless, as Pascal's 'Memorial', from which the title of this book is taken, suggests, there are things about God that science cannot know. I shall argue that the 'God of the men of science' is the same God as the 'God of Abraham', but that religious faith requires a distinctive personal experience of God that science cannot give. That requirement draws attention to important areas of personal experience with which the sciences do not deal, but which are an essential part of any adequate account of the nature of reality. So the central intellectual problem of this book is the relation between scientific enquiry – which presupposes faith in the intelligibility of nature – and personal experience of reality – which inspires, for many people, a search for deeper religious understanding.

I have held chairs of philosophy and theology. So it is obvious to me that human beings agree on very little and that there are at least two sides to almost every argument. In fact, philosophy, as taught in Britain, often induces a sort of permanent intellectual paralysis, because you can always see the objections to any belief you may be tempted to have.

Nevertheless, in morality, in politics, in history, in art and in religion, too, we do have some fundamental beliefs, which we will not give up unless we meet objections that seem absolutely overwhelming. In that way, I am committed to belief in God, as the most morally demanding, psychologically enriching, intellectually satisfying and imaginatively fruitful hypothesis about the ultimate nature of reality known to me. That is why the discoveries of modern science are important to me. If science, as Professor Dawkins and some other scientists claim, somehow undermines the rationality of belief in God, that would provoke an intellectual crisis for me. But it would just have to be faced and, as a properly trained philosopher, I am quite prepared to face it.

It may be, however, that the sciences, in giving new knowledge about the world, do not undermine belief in God, but might arguably support it. They might nevertheless change what can reasonably be thought about God, about what sort of God there might be and about how we believers are to think of God in a scientific age.

I have been very lucky in being able to meet a great many leading scientists and to discuss these matters with them. I know I will never experience the thrill of discovering a new scientific fact or propounding a new scientific theory. On the other hand, I have read more books about God than most other people, and I have even written more books about God than most other people have read. I have contributed to books whose main contributors were leading scientists, and which represent the discussions we shared together (I have listed these articles of mine at the end of this book). So I have tried to acquaint myself with the best current scientific thinking and apply it, wherever relevant, to the idea of God, the creator of this marvelous and beautiful universe.

I intend to show what the consequences of knowledge of modern science can be for our thought about God, and why such knowledge is of very great importance for the future of religion. Indeed, I hope to show that some knowledge of modern science is necessary for an educated and reasonable belief in God, that such knowledge is accessible and should be part of any reputable religious education. Nevertheless, the heart of religion is personal experience of that transcendent reality towards which much science points, and so the challenge for religious believers is to integrate such personal life-transforming experience with the best available scientific knowledge. Since such knowledge is constantly changing, this book will regrettably be out of date before it is published. I can only plead that I checked it with reputable scientific authorities as I wrote it, and it may stand at least as a stage in that continuing process of seeking deeper understanding which will always characterise mature religious faith. The suggestions for reading at the end of the book are meant to provide accessible and reliable introductions for those with no specialist knowledge.

INTRODUCTION

God was declared dead in 1883 by Friedrich Nietzsche. God may have been killed by many things, but a major suspect was science. Battles over the literal truth of the Book of Genesis and its account of creation, over the occurrence of miracles, over evolution and over the causation of disease by demons have given the impression that science and religion are deeply opposed to one another. Moreover, scientific demands for repeated experiment, detailed evidence and provisional hypotheses all seem to be ignored by religion, which apparently asks for faith and absolute certainty on very little evidence at all.

In the opinion of most people, science has won those battles, and God should by now really be quite dead. But something rather unexpected has happened. Believers and theologians have been reduced to a defensive silence about God, and are rather apologetic about even mentioning the word. But scientists, especially theoretical physicists, have started discussing God again, though from a rather different point of view.

If you walk into any good bookshop, prominently on display will be many popular books on science by reputable scientists. It has become quite usual for these books to contain a final chapter on God. Questions are seriously raised: could God be the final explanation of everything? Does the apparently rational and highly organised structure of the universe suggest a vast cosmic intelligence? Does the universe require an origin beyond itself?

A good number of scientists believe that some sort of cosmic intelligence must lie behind the incredible complexity and order of the universe. They often add that they are not talking about the God of religion (so Paul Davies writes, 'Science offers a surer path than religion in the search of God': *God and the New Physics*, Penguin, Harmondsworth, 1983, p. 229). Some of them still see the religious God as some sort of tyrant who orders people around, laying down irrational laws that are never to be questioned, and requiring everyone to keep telling him how great he is and to kill everybody who disagrees. But they are talking about a being of consciousness, of awesome intelligence and power, which sets the universe up in accordance with elegant mathematical principles. This is a God, even if it does not ever interfere in the universe, or demand to be worshipped.

Of course not all scientists agree. Some – a small but very vocal minority – think it is a mistake to talk about God at all. They argue that proper science shows the universe to be purely material. Intelligence is thrown up in the course of evolution as a sort of by-product, and could not exist without any physical basis, or on its own even without any physical universe, as it would have to do if there was a God.

Over a century after God was declared defunct, the argument still rages, and recently with renewed force. Does science show the universe to be completely physical, so that God never existed in the first place? Or does it show the universe to be a much more mysterious place than materialists think, so that God as a cosmic consciousness and intelligence is a highly plausible postulate for explaining the form and order of the cosmos? Or does it even support a more religious God, a personal creator who desires the love of creatures and has a positive purpose in creating the universe?

These are the questions with which this book is concerned. Naturally, I have a view of my own. After twenty-five years of university teaching in philosophy and fifteen years of teaching theology, I am greatly struck by the way that modern physics has put God back on the intellectual agenda. When I started studying philosophy, God was not on that agenda – the philosopher A.J. Ayer once said that he was not an atheist, because it was logically impossible to deny something that was devoid of meaning.

Metaphysics, as well as God, was clinically dead. But, like God, it has been resuscitated – by scientists who make no bones about telling us what reality is really like. Like the metaphysicians of old, they disagree with each other, but at least they think they are disagreeing about something. They are not just speaking various dialects of nonsense.

Not only is God back on the agenda, a great deal of modern science – some of it developed only at the very end of the twentieth century – can be seen as positively pointing to the existence of God, in the sense of a cosmic intelligence, as the ultimate basis of physical reality.

I shall try to show just how this is so, and how very extensive and varied the range of contemporary scientific data is that points to the existence of ultimate mind. But all assertions in this area are controversial, and I shall also try to be honest over the key issues about which some intellectual differences are bound to arise. Finally, I want to convince friendly scientists that the scientific view of an ultimate cosmic intelligence is much more like the religious God than they may think. To do that, I will have to convince friendly believers in God that their religious God needs to be put much more firmly in a scientific context if it is properly to come to life.

The book is in three parts. In the first part I outline the major elements of the revolution in worldview that the sciences have brought about. I show how the new scientific worldview has given rise to a purely naturalistic account of a cosmos that is purely physical, but has also given rise to an alternative account of the cosmos as caused by a super-intelligent mind, which could be called God. In the second part, I examine the claim, made by some scientists, that naturalistic science can provide an 'explanation of everything', and ask what such an explanation might properly be like. I argue that God might provide a more satisfactory explanation, and outline what such an explanation might be like. In the third part, I show what a more overtly religious idea of God adds to such an account. I ask what sort of ultimate goal religion can posit for the universe and for human life. And I suggest how the scientific view might reasonably lead on to a religious view, and how belief in a religious God might properly be influenced by contemporary science.

Traditional religion will need to change in many ways if it is to cease waging war with modern science. But why should only religion remain the same for ever, while science is continually advancing to new insights and discoveries? Extending knowledge, in both religion and science, is one of the joys of human life. It is important to this process that each person who can should pursue their own path of discovery, knowing what has been said before them. So, though I will be defending my view that ultimate mind is indeed the basis of physical reality, my aim is primarily that readers should perhaps see the issues in a new way and enjoy making their own intellectual decisions and discoveries.

PART I

THE FORMATION OF THE SCIENTIFIC WORLDVIEW

1

THE END OF THE ANTHROPOCENTRIC UNIVERSE

In many traditional religious views, human beings are the most important things in the universe, and the whole of nature is created to serve humans. After Galileo, that view was turned completely on its head. Nature is created to express the glory of God. The role of humans is to conserve and shape nature and possibly to share in the creative power of God and the richness of God's awareness of the cosmos. Humans are one among possibly many forms, and not the final form, of intelligent life. But humans need not be seen as accidental or peripheral to the universe – they can be seen as an integral, if small, part of the divine plan.

AIM: To show that the conflict between Galileo and the Roman Catholic Church was not primarily a conflict between 'religion' and 'science'. To show that parts of the Bible should not be taken literally, especially when they deal with events for which the writers had little or no evidence. To show that it magnifies God's glory to know that humans are small parts of a much greater universe, which God values for its own sake.

1. GALILEO AND ARISTOTLE

The rise of the natural sciences has changed for ever the way we look at the universe. In fact it has changed our view of the universe not once, but at least four times. Each time there was a conflict with more traditional theories that had been held until then, but on each occasion the new view was victorious.

Some historians talk as though the conflicts were between science and religion, and this has generated the myth that science and religion are bound to conflict and that science always wins while religion always retreats. This is a distorted view of what actually happened. The conflict on each occasion was between traditional science and new science, and there were religious believers on both sides of the conflict every time.

In the first part of this book I shall look at four major changes of outlook that the sciences brought about. The first three can be associated quite easily with specific big names of science, and it would not be hard for anyone to guess who they are going to be. Galileo, Newton and Darwin would be on anyone's list of the greatest scientists of all time. The fourth is more difficult. You might expect the big name to be Einstein, and in a way it was. Yet the fourth change was brought about by quantum theory, and Einstein was actually opposed to many of the ways in which quantum theory is now usually interpreted. So, surprisingly, he represents the end of the old physics more than the birth of the new (quantum) physics.

The first big change, though, has to be associated with the name of Galileo. The legend has grown up that this was the first and decisive battle between science and religion. Galileo certainly won the battle, so some popular histories of science depict the story as the beginning of the death of God and the triumph of materialism.

Nothing could be further from the truth. Galileo was, and always remained, a devout Catholic. There are many readily available scholarly accounts of the dispute between Galileo and the Church, which show the error of depicting it as a battle between progressive science and reactionary religion. James Reston, Jr's

Galileo: A Life (HarperCollins, New York, 1994) is just one work that puts the dispute in its proper context, and Richard Blackwell, in *Galileo, Bellarmine and the Bible* (University of Notre Dame Press, Notre Dame, 1991), focuses on the religious dimensions of the debate.

There was certainly a conflict between Galileo and the Roman Catholic Church, which came to a head in 1633 when Galileo was condemned by the Holy Inquisition and forced to retract (or say that he retracted) his Copernican views.

The Polish Catholic lay canon and astronomer Nicolaus Copernicus had asserted that the earth circled the sun in *On the Revolutions of the Heavenly Spheres*, published in 1543, the year of Copernicus's death, but there had been little public reaction. Copernicus's book was dedicated to Pope Paul III and carried the endorsement of the author's local cardinal. However, in 1616 the consultants to the Congregation of the Holy Office (the Inquisition) declared that the heliocentric hypothesis was formally heretical. And when Galileo later re-affirmed the Copernican hypothesis in a very combative way, an affronted scientific establishment took action. Galileo was convicted in 1633 of disobeying an injunction allegedly placed on him in 1616 not to promulgate Copernican ideas. He was placed under house arrest, mostly in his villa near Florence, until his death nine years later.

The conflict, however, was not so much between Christian faith and the Copernican view that the earth circles the sun, as between established Aristotelian science and the 'new science' of close observation and experiment that was threatening the old scientific elite.

The Catholic Church had associated itself firmly with the authority of Aristotle, who was taken to be master of all sciences (except theology, where he needed to be corrected by Thomas Aquinas, the 'angelic doctor'). Aristotle's concepts of substance and accident, form and matter, act and potency had been used in framing doctrines like that of transubstantiation and ideas of God. His system of the four types of causality, material, formal, efficient and final, was accepted as the proper framework for natural science. Partly because of this, his physics was accepted as definitive, and the Biblical account of the universe was largely interpreted in terms of it.

Ironically, the medieval use of Aristotle was itself a major revision of the Bible, since the Hebrew, Biblical, idea of the cosmos was not very like Aristotle's. Nevertheless, Aristotle, after some initial suspicion (the Archbishop of Paris prohibited the teaching of Aristotle in the thirteenth century), had become identified with the dogmatic definitions of the Catholic Church, and with the scientific establishment, which was sponsored by the Catholic Church. Galileo's insistence on close observation with instruments like the newly invented telescope, and on repeated experiment, seemed to threaten accepted scientific orthodoxy, and it was strongly resisted by the guardians of that orthodoxy, which was entrenched in the institutions of the Church.

2. FAITH AND FACT

No one doubts today that Galileo was correct. In 1992 the Catholic Church formally rehabilitated him, thereby admitting that the Inquisition had been in error. Matters of scientific fact are not matters of faith, and if an authoritative interpretation of Scripture includes matters of scientific fact it may be, and it demonstrably has been, in error.

This is a point of great importance for religion. Many, probably most, alleged religious revelations include or presuppose statements about matters of scientific fact, about the nature of the cosmos and the place of humanity within it. If all such statements are fallible and some of them are mistaken, it becomes very important to decide exactly which matters are the province of the natural sciences and which the province of revealed faith. At one time it was possible to say that everything in a holy text was guaranteed true, but after Galileo we must first ask what is the subject matter of the text. Only matters of faith, not of science, can be said to be immune from error. But how are we to distinguish the two? Or, if we cannot do so, what are we to make of a faith that is prone to error? Can we any longer trust it?

One move that we might make is to deny that the Bible, which was the text in question in the Galileo case, contains any statements of scientific fact, and to say that the mistake the Inquisition made was to think that it did. The Church had long been used to interpreting Biblical statements about God metaphorically. God does not literally trample on the nations (Habakkuk 3:12). So it is not too big a move to take apparently factual statements about the physical universe metaphorically. Saying that it was created in six days, for instance, had usually been taken metaphorically. 'Days' were normally thought to be not periods of twenty-four hours, but possibly vast periods of time. So it is not too hard to take the whole account metaphorically, and say that the point of the Genesis stories is not to tell how the universe originated, but to put in story form some important spiritual truths about the relation of humans to God and their responsibilities to creation.

Cardinal Bellarmine, writing in 1616 to Galileo's friend Foscarini, saw this possibility, though it was not considered by the authorities that Galileo had yet proved his case. Bellarmine writes:

> I say that if a real *proof* be found that the sun is fixed and does not revolve round the earth, but the earth round the sun, then it will be necessary, very carefully, to proceed to the explanation of the passages of *Scripture* which appear to be contrary, and we should rather say that we have misunderstood these than pronounce that to be false which is demonstrated.

The Catholic Church has now made this move, and officially and unequivocally accepts a scientific account of the origin of the universe. But does this have implications for faith in God? It does entail a denial of Biblical literalism, and it does show the importance of metaphor and story in the Bible (and in any other religious text). That is important for particular religions like Christianity, Judaism and Islam, but it does not affect belief in God as such.

There is no doubt, however, that a Galilean view of the universe differs from that of most traditional religious views. Before the sixteenth century in the West there was a generally held Christian belief that the created universe had only existed for a few thousand years, that it all existed for the sake of human beings and that it could not be expected to exist for very much longer.

In India there was a very different view that the universe had existed for innumerable aeons, and that it was only one of an infinite number of universes, so that humans were a very small part of finite reality. Yet the universe was thought to be morally ordered by the laws of reincarnation and of karma, according to which all acts receive their due deserts sooner or later. Many of the things that happen are due to the actions of gods or spirits, whose presence is often apparent in the physical universe. The present age is a degradation from an earlier golden age, when the gods walked the earth and when humans were more intelligent and less self-centred.

The discoveries that the sciences have made about the universe put these, and all other pre-scientific cosmologies, in question. There is a huge contrast between Aristotle's view of the stars as set on a crystalline sphere above the spheres of the sun and moon, and what everyone now should know, that our sun is just one star among billions in the galaxy of the Milky Way, which is itself just one among billions of galaxies in this cosmos, which may also be just one among billions of universes, possibly generated within black holes.

The information about the universe the sciences have uncovered is enough to change radically all pre-scientific theories of what the universe is like. To put it bluntly, all pre-scientific views of the physical universe were incorrect in at least some important respects. To the extent that ancient religious and philosophical beliefs incorporate references to the nature of the physical universe, they stand in need of revision. How far must such revision extend, and how important will such revision be for religious traditions originating long before the age of science?

This is the problem the Galilean dispute bequeathed to the Christian Church, and by implication to all religions.

3. HUMANS REMOVED FROM THE CENTRE
OF THE UNIVERSE

Apart from this problem, Galileo's astronomical theories do not seem to pose a threat to religious belief as such. But they do suggest the need for a fairly drastic revision of the traditional Christian worldview.

The traditional view was that God had created the earth as the centre of the universe, and placed human beings uniquely at the apex of the created order, so that everything in creation ultimately existed 'for the sake of man', as Thomas Aquinas put it – 'The whole of material nature exists for man' (cited in Paul Santmire, *The Travail of Nature*, Fortress, Philadelphia, 1985, p. 91).

When in 1608 Galileo heard that a Dutch lens grinder, Hans Lippershey, had made a refracting telescope, he immediately set out to construct a telescope himself. Using this, the first astronomical telescope, Galileo was able to show that the Milky Way consisted of distant stars, that Jupiter had moons and that Venus had phases, which was best explained by supposing that it circled a central sun.

From these beginnings, our ability to probe space has increased enormously. The Hubble Space Telescope, launched in 1990, enables us to receive infrared wavelengths from the oldest objects in the universe, which existed billions of years ago. Galileo's observations, though they could hardly have envisaged the vast universe we can now see, laid the foundation for our realisation that humans are not likely to be the centre of God's purposes, since the universe is much vaster than was ever imagined and the earth is not the centre of it.

It does not follow that, just because we are not physically at the centre of the universe, we are not central to God's plans. This planet may still possibly be the only inhabited world in the whole universe. At the time of writing, no communication from any other inhabited planet has been received. As the physicist Enrico Fermi asked, 'Why aren't the aliens here?' Since some stars are billions of years older than the sun, we might have expected there to be some intelligent life in the universe long before we existed. Why has it not communicated with us? If there is no intelligent life anywhere else in the universe, it might be said that conscious intelligent agents are of more value than billions of light years of unconscious and unintelligent space. Humans could still be the culminating point of the universe, even if the earth is not physically at the centre of it.

In that case, Galileo's relocation of the planet earth would have no religious implications at all. It would not matter where we were located physically, as long as we were the central focus of God's love. But a great many astronomers, following up Galileo's relocation of the planet earth in the universe with more recent discoveries about the vastness of space, see things differently. They make much of the fact that humans have only existed for a tiny fraction of the history of the cosmos, which is thought to have originated with the big bang between thirteen and twenty thousand million years ago. I am trying to be careful about this number, even though the current favourite is fourteen billion years (that is an American 'billion', which is much smaller than a British 'billion'). I am very conscious that when you quote a number like this, some physicist will say, with a pitying smile, 'Ah, that was last week's theory.' Still, it seems to be agreed that it all started a very long time ago.

In addition to this, the conditions of human existence are so very precarious and possibly accidental (dependent on, for instance, the extinction of the dinosaurs sixty-five million years ago, and on the planet not being hit by a major comet or asteroid since) that it does not seem that they could be planned at all. Moreover, humans are doomed to become extinct, since the sun will die in five billion years and the earth will become uninhabitable. Even if we survive that by travelling far into space, the whole universe, in accordance with the laws of thermodynamics, will inevitably become uninhabitable eventually. At least, it will be uninhabitable by any form of life remotely like ours. (See the discussion in chapter 17.)

Predictions of the ultimate end of the universe change with great rapidity, and there is a danger that by the time you read this sentence quite a new theory will have been devised. Yet it seems agreed by cosmologists that conditions in the far future universe will be quite different from how they are now. The present view (in June 2006) is that the universe will continue expanding, and perhaps accelerating (see Martin Rees, *Our Cosmic Habitat*, Princeton University Press, Princeton, 2001, chapter 8). Eventually, all stars and planets will die, groups of galaxies will merge and we will be left with 'a single amorphous system of aging stars and dark matter' (p. 118). As Paul Davies puts it, 'In a standard picture the very far future of the universe would be characterised by a dilute and ever-diminishing soup of extremely low-energy photons, neutrinos, and gravitons moving virtually freely through a slowly expanding space. Normal matter, if it exists at all, might consist of the occasional electron and positron drifting slowly through this invisible soup' (Paul Davies, 'Eternity', in *The Far-Future Universe*, ed. George Ellis, Templeton Foundation Press, Radnor, PA, 2002, p. 47). The prospect does not seem too enticing, and it does not seem that anything like human beings could exist at that point. Humans are inevitably destined for death, and that may seem to reduce all human achievement and purpose to a pointless blip in the story of a pointless universe. In the light of these facts, it can seem implausible to think the whole universe exists just for the sake of such short-lived, accidental and doomed organisms as we are.

Galilean thought, or more correctly this cosmic perspective that is the heir of Galilean thought, can suggest the dethronement of humanity from any important place in the cosmos, and the lack of any objective purpose in human existence. It certainly undermines the traditional worldview of Judaism, Christianity and Islam.

4. THE BEAUTY OF THE COSMOS

But on closer reflection it seems more reasonable to say that the post-Galilean picture of the universe calls for a revision, indeed an enlargement, of the traditional religious worldview, rather than for its complete renunciation. For anyone who believes in a creator God can affirm that the cosmos is created so that God can enjoy its beauty. After all, theists believe that the cosmos is a product of the divine mind, so its creation can be compared to the work of a supreme artist, enjoyable and worthwhile for its own sake, without any reference to possible finite persons at all. It would not matter if there were never any human beings at all. The universe could still have a point, and that point would be its expression of the power and wisdom of the creator, and God's enjoyment both of the process of creating and of the created universe itself. That is part of the traditional view – the Hebrew Bible depicts the divine Wisdom as 'rejoicing in his [God's] inhabited world, and delighting in the human race' (Proverbs 8:30). And if Wisdom delights in the human race, it surely also delights in the beauty of the stars. If there is a God, the universe has a point, as the creative expression of the master creator, and the object of divine delight.

Still, if finite persons come into being, they could share, to some small degree, in that creation and appreciation of beauty. The theologian Thomas Aquinas said that 'Goodness seeks to share itself.' So, if God's experience of the universe is good, God might desire that it should be shared with finite persons. Traditionally, only human beings are considered as finite persons. In the New Testament, it is said that God promises that humans 'may become participants in the divine nature' (2 Peter 1:4). So traditional Christianity certainly thinks of humans being created so that they can share God's experience and rejoice in the wonder of the universe. And I think that is true, in slightly different ways, of traditional Judaism and Islam too.

When we realise that humans might not be the only conscious rational personal beings in the universe, this does not undermine the traditional view. It extends it to cover whatever sorts of finite persons there might be, in different galaxies and living in the light of different stars. It does not really matter how many different sorts of persons there are, or for how long they exist, long or short, late or soon. It is their capacity for the appreciation of the beauty of being, for understanding the nature of the cosmos and for creating new forms of beauty themselves which makes them special and gives an additional meaning to the universe, as a place where created persons can share the divine delight in creation for its beauty and wisdom.

Far from undermining religious views, it is a genuine new religious insight that humans are only tiny parts of creation and that the universe was not created just to serve us. Perhaps we were created to serve the universe by enjoying, conserving and shaping it to some extent. Humans may still have a distinctive role, in our capacity to understand and act in the universe, as members of a possibly much larger class of conscious agents who exist in the universe.

5. ARE HUMANS ACCIDENTAL?

Modern science seems to suggest that the existence of human persons is a freakish accident. Or does it? If there is a God, and if God wills humans to exist, then human existence will not be an accident. The laws will be set up so that we will come into being, however improbable it seems to us. What looks like a lucky accident, in a world without God, will be seen as a providential arrangement if there is a God. The only question to ask is: does it look as though a God could have intended to produce us? The main argument against that perspective is that humans seem to exist because of a long chain of highly improbable lucky breaks. Our planet needs to be just the right distance from the sun to prevent it being too cold or too hot for life to develop. It needs to be protected by large outer planets from comets that would destroy it. It needs to have just the right amount of oxygen, carbon and hydrogen to support life. The list of conditions goes on and on – Martin Rees's book spells out more of these details. If each step in the chain is pretty unlikely, the whole chain is immensely improbable.

There are two ways of looking at that. Either you can say, 'The fact that we exist is a freaky accident, and it is unlikely that there is any other life in the universe at all' – that is basically the view of Stephen J. Gould. Or you could say, 'This process is so improbable that it looks as if the whole thing has been fixed.

When you get as many unlikely coincidences as that, you cannot really believe it is an accident.' As the mathematical physicist Paul Davies says, 'I cannot believe that our existence in this universe is a mere quirk of fate ... we are truly meant to be here' (*The Mind of God*, Simon & Schuster, New York, 1992, p. 242).

We can perhaps all agree that the existence of human life is highly improbable. But that does not show that a God did not intend it and make it happen. The probability of an event is always relative to some background information, to some set of facts against which probability can be measured. To assess how improbable an event is, we need to know what possible alternative events could have happened. When Stephen Gould says that it is highly improbable that the earth should have survived the impact of gigantic comets, he means that, in accordance with the laws of nature, there are many chances that comets should hit the earth – and, of course, many do, though not quite enough to have destroyed all human life (yet). If the background facts are just the laws of nature, they seem to allow for many possibilities of disaster – and the more such possibilities there are, the more improbable our survival will be.

Suppose, however, that we allow God, an intelligent mind that wishes intelligent life-forms to exist, as a background fact. That will affect considerably the probability of such life-forms coming to exist. In fact it will make the existence of some of them certain – having a probability of 1. Without a God, the existence of humans is very very improbable. But with a God, it is very highly probable (not quite certain, since humans might not be quite what God wants, in the end).

It is a good scientific principle to accept a hypothesis that makes some state of affairs more probable than it would be without the hypothesis. So it looks as though a state of affairs that is very improbable without God, but very probable with God, suggests that the hypothesis of God is a reasonable one. The freaky good luck that Gould points to in fact makes it more likely that some super-intelligence is behind it.

Of course it is not as simple as that. People may doubt whether the existence of human life is really valuable enough to be a worthwhile aim for God. Was it really worth billions of years of cosmic evolution just to produce Tony Blair? (I have nothing against Tony Blair, but some people seem to doubt that he could be the final purpose of the universe.)

People may also have problems with the existence of an immaterial being. 'Where could it be?' they may ask, and remain unsatisfied with the answer that it is not anywhere at all. And people may object to bringing in a scientifically unknown entity to help explain why observed facts are the way they are. Can God, who is admitted to be a supreme mystery, really explain what is only tremendously puzzling?

There are some problems with using God as a good explanation of the universe. Nevertheless, we cannot say that science *shows* that humans are nothing but freakish accidents in the universe. What it shows is that a remarkably large number of improbable states – more than scientists in the past would ever have expected – have to exist for us to be here. Either we are extremely lucky or the process has been fixed.

We cannot tell from science whether humans are here by chance or not. But all the facts are compatible with the laws of nature being 'fixed' so that humans are bound to come into existence. Some super-intelligence would be needed to do the fixing. Post-Galilean science does not, after all, show that humans are an accident in the universe. On the contrary, to scientists like Paul Davies it suggests the opposite – that the existence of humans, and perhaps of other intelligent life-forms too, was built into the universe from the start. Perhaps we are not accidental. Perhaps we are, as he says, 'meant to be here', implicit in and intended by the universe or its creator from the first.

6. ARE HUMANS DOOMED?

A pessimistic reading of post-Galilean science is supposed to make us all depressed because, at least on one standard reading, all life, all stars and the universe itself will one day become extinct. Yet that is no more than saying what we all know, that we will all die anyway, most of us long before that. I suppose some people might get depressed at the thought that they are going to die. But is it really reasonable to say, 'I am so depressed. It's all going to end'? 'When?' 'Probably in many billions of years.' It might seem not worth worrying about too much just yet. It might be more reasonable to say, 'I have only got a few years of life, so I am going to make the most of it, enjoy it as much as possible, and put into it all that I can to make it better.'

If someone says, 'Let's visit Venice, one of the most beautiful cities in the world,' is it rational to say, 'There's no point; we are all going to die'? I do not think so. A short life can be filled with beauty, with the pursuit of truth, with friendship and happiness. And we might even enjoy a trip to Venice more intensely if we know we are seeing it for the last and only time. So the thought of the ultimate extinction of the universe does not mean there is no point in it. It does not mean there is no purpose in it, since the purpose could be in the process itself, not some last state at which it aims.

One purpose of the universe could well be to generate finite persons who do not live very long but who can accomplish many valuable things and enjoy many valuable experiences in their short lives. There is, of course, the problem that they might have some bad experiences too. That is the biggest objection to the whole process being created by a God, and it is a very big objection. Perhaps that is why some scientists will accept the existence of a super-intelligence but remain sceptical about whether it is good or benevolent, whether it is the religious God.

Even if there is no adequate response to that objection, still the fact the universe is ultimately doomed does not deprive it of purpose or meaning. It could be created for the sake of all the good things it contains, even if those things do not last for ever.

If there really is a God, then in a sense all good things will last for ever. They will exist in the mind of God. Almost everyone who believes in God believes that God experiences all things and is aware of all that happens. Moreover, God's knowledge is not subject to any deficiency. God will never forget anything – that is part of what it means to say that God is omniscient or all-knowing. If that is the case, then all the good things that have ever existed in the universe will continue to exist in the mind of God after the universe itself has ended. But that could not happen unless the universe first existed, to produce lots of values that would otherwise never have existed at all.

The only drawback is that God will presumably remember all the bad things as well. This is a much less entrancing prospect, not least for God. For that reason, theists have often held that God has some way of transforming the bad things, or of making them less intense and destructive, perhaps even of suppressing them from consciousness ('The former things shall not be remembered or come to mind': Isaiah 65:17). God has a way of setting them in a wider context that shows them to be necessary parts of a process that produces many good things. So, in the divine mind, the evil of suffering and pain is transformed by being included in a much wider process that brings about overwhelmingly greater good.

It is notoriously difficult to spell this out in detail, but it is part of the religious vision of the Hebrew prophets, who look for a time when 'The Lord God will wipe away the tears from all faces' and 'swallow up death for ever' (Isaiah 25:8). This implies that, not only will God remember all the good things that have ever been in the universe, and see the evil things in that wider context of good, but in some way God will give finite persons some share in this divine mind and experience. This would be a form of life after death, or perhaps of life beyond the confines of space and time. It would help to lessen the problem of the suffering in creation. But for the Hebrew prophets this is a sort of bonus, not a necessary condition of the cosmos having a purpose and value. Belief in personal immortality was not clearly formulated in ancient prophetic tradition. The prophets were more concerned to say that the universe has eternal value because it is always held in the mind of God, and within the universe humans may play an important, if not quite a central, part in making the universe what it is.

Once you start talking about life after death, you may seem to have gone well beyond the boundaries of science, though in chapters 16 and 17 I will show that science has some surprising suggestions to make. But even if you discount all talk about life after death, our scientific knowledge of the universe does not undermine a belief that the universe has a purpose, and that it is created by a wise God.

7. THE VALUE OF THE COSMOS

So the hugely expanded view of the universe that started with Galileo's telescope does not deprive humans of their significance, or the universe of purpose and value. The existence of God may not be strictly necessary for human life to have significance, or for the universe to be of great value. But it would ensure that the universe is of great value, a value that is not just invented by human minds. For God, according to religious traditions, creates a universe precisely because it is of value, and so the universe has that value, whether humans think so or not. Human lives too will have an added significance if there is a God, for they will be of value to God and will have a part to play in God's creative purposes.

One of the values of the universe is that it is supremely intelligible and mathematically beautiful. One of the purposes of human life is that, by the use of our intellects, we can come to understand and enjoy that intelligibility and beauty. Because of this, belief in God actually encourages the pursuit of scientific truth and ought to be a strongly motivating force to pursue such truth.

The scientific worldview therefore does not have to take the pessimistic view that humans are insignificant parts of a vast, impersonal, accidental and pointless universe that is doomed to extinction. It is equally compatible with the much more positive view that the universe is a beautiful and elegant expression of a supreme cosmic intelligence, and that it has a purpose, which is simply to actualise states and processes of intrinsic value. Humans have a small but significant role to play in knowing and appreciating these values in new ways, and perhaps even in creating new forms of value in their part of the cosmos. So they can add new forms of actual value to the universe. However small and however brief those values are, they have a genuine importance of their own.

This is a religious vision of the universe, and not all scientists will accept it. But far from religion and science coming into conflict in the time of Galileo, it is clear that the scientific views of Galileo, though opposed by the entrenched scientific establishment of his day, are positively related to belief in a creator God. It is not surprising that modern science largely originated in the work of deeply religious men who understood their work as sharing in the wisdom of the creator and who often dedicated their work to the glory of God.

2

THE INTELLECTUAL
BEAUTY OF BEING

The Newtonian revolution was to see nature as the intelligible product of one rational and elegant cosmic mind. This later gave rise to the idea of nature as an impersonal machine, whose laws are absolute, inflexible and all-explaining. For Newton himself, however, the laws of nature demonstrate the presence and power of one supreme God of immense wisdom, intelligence and spiritual purpose. For Newton, science is a spiritual enterprise, seeking greater understanding of the wisdom of God.

AIM: To show that much early modern science was prompted by belief in one wise, rational creator God. To show that the 'laws of nature' do not exclude the actions of God, but show God to be a wise, rational and powerful creator.

1. NEWTON AND THE LAWS OF NATURE

The second great change in human understanding of the cosmos was made above all by Isaac Newton, whose *Principia*

Mathematica, published in 1687, set out the basic laws of mechanics and motion and consolidated the scientific perception of the universe as governed by universal laws of nature. Newton, like Johannes Kepler before him, was a devout, if technically unorthodox, Christian, and saw his discovery of the simple, elegant and intelligible laws of nature as an appreciation of the wisdom of the divine creator.

The idea that nature operates in accordance with general laws, so regular and precise that they can be captured in mathematical equations, is astonishing and unexpected. For most of human history events in nature were widely believed to be caused by the actions of spirits, gods or demons. Their actions often seemed to be arbitrary, or contradictory. Even if there is only one supreme God, God might do anything at any time. God's ways are inscrutable, so it might even be impious to examine the modes of God's operations in nature.

Newton's religious belief was that there is just one God who is supremely wise or rational, and therefore not arbitrary, and that humans are made in the image of God, so that they can in principle understand the rational principles on which God would create the universe. The hypothesis that the universe proceeds in accordance with intelligible laws, which will be beautiful, elegant and reliable, though not invented by Newton, was confirmed in a definitive way by him.

2. THE DISENCHANTMENT OF NATURE

Though it was a reflection of Newton's own religious faith, this view of the universe as governed by absolute and inviolable laws did threaten many traditional religious beliefs that were widely held in the seventeenth century. Newton himself had no difficulty in believing in miracles, since God could break the laws God had made whenever God chose. But some of Newton's followers, convinced that God was supremely rational, concluded that God would have created the most efficient system possible, so that miracles (conceived as violations of laws) would be unnecessary and somehow unfitting. Surely a rational God could have designed the universe so that miracles were unnecessary?

So it has almost become part of the modern scientific world-view that miracles do not occur. I say 'almost', because no science can show that the laws of nature are never broken. After all, they may be broken when there are no scientific observers present. Still, it may seem suspicious that miracles have never been scientifically observed under experimentally repeatable conditions and that they do not seem necessary to explain the natural world successfully. So it has become part of the methodology of science that we should always seek a naturalistic explanation of events, one that does not appeal to supernatural or divine causality.

The best example of this is perhaps in medicine, where belief that demons cause illness has been replaced by belief in viruses and antibiotics. The natural causes of illness can be counteracted much more effectively by medicine than by exorcism, just as plant diseases can be more effectively treated by insecticides than by dancing naked by the light of the new moon. Very few people doubt that disease should be eradicated by medical treatment and not meekly accepted as the result of demonic possession or as a punishment for sin.

A much better understanding of the world is obtained when we can identify the causal laws in accordance with which natural processes occur than when we look for some supernatural reason why things happen. In a similar way, it has proved much more effective to ask what the general causes of events are than to look for some reason why they happen.

Just as Galilean science drove out the Aristotelian view of astronomy, so Newtonian science drives out the Aristotelian search for final causes – for some purpose for the sake of which specific things happen. If someone falls ill, we no longer ask, 'Why [for what reason or for the sake of what goal] did this happen to this person?' We do not think that if someone gets mumps, God is punishing them for not going to church. Instead we ask what natural process caused it to happen, and then we seek to counteract that process by the use of chemicals or surgery. We intervene in nature to improve it, and to do that we have to understand it as a set of events causally connected by laws that we can discover by observation and experiment.

The naturalistic methodology of science has proved very effective. Human life has changed immeasurably for the better since the discovery of anaesthetics. Nature can be understood in terms simply of the causal laws of its own operation, and then it can be improved. So nature is no longer seen as set up in detail by God so that good acts are rewarded and bad ones punished, and whatever nature does is simply to be accepted with resignation. The laws of nature are morally neutral, and we can use them to make human life better.

This has been called the 'disenchantment of nature', since it involves thinking that nature is no longer an inviolable sphere of divine institution, something sacred and untouchable. It is more like a mechanism – Descartes spoke of it as a machine – that can be taken apart and re-assembled to make it more efficient.

Seeing the whole cosmos in terms of this machine metaphor is not necessarily implied by Newtonian science. But it is a major part of what is often understood as the scientific worldview to think that there are no final causes, no purposes, in nature. All events are inevitably caused by prior physical states plus a set of universal laws, and technology can in principle and should try to make the human condition better by responsibly controlling the processes of nature (particularly to eliminate suffering and prolong life).

Within such a worldview, talk of particular divine providence and action can come to seem at best odd and at worst irrational. Perhaps a supremely wise designer set up the whole system, but then the universe proceeds without further reference to the acts and intentions of God. It also becomes difficult to believe that nature is in order as it is, so that it would be wrong to interfere with or improve it. Nature, however astonishingly rational and beautiful, is morally neutral and it is not perfect – it can be improved and much suffering and pain can be eliminated. So humans have a more creative and responsible part to play in the improvement of nature than many traditional believers had thought.

3. IMMANUEL KANT, FREEDOM AND DETERMINISM

One possible religious consequence of such a view can be seen in the work of Immanuel Kant (1724–1804). Influenced by his understanding of Newtonian science, Kant believed that every event in nature had a sufficient or determining cause. That is, given its antecedent physical state and the laws of nature, each event happens by necessity; there is no alternative to it, and it has to be just what it is. This is the thesis of physical determinism, which might be called the scientific reformulation of Calvin's hypothesis of total predestination by God.

For Kant, this entailed that there can be no miracles, no supernatural interventions in a physical order that is a closed and complete causal system. There can also be no experiences of God. For if I experience something, that thing is at least a part cause of my experience of it. If there can be no supernatural causes in nature, it follows that no one can actually have experiences of a God, who is a supernatural being. Prayers to such a supernatural being will obviously have no effect on the world. And as for Christian doctrines like the incarnation or the divine presence in the sacraments, these are impossible. The supernatural and the natural do not mix.

But Kant was also influenced by the new stress on human creativity and responsibility for improving the world that science-based technology made possible. Moral freedom and human autonomy became fundamental values for him. This created a major problem. How is moral freedom compatible with physical determinism? Many philosophers have wrestled with this problem, and come up with very different answers. Kant himself was forced into the desperate expedient of saying that I am morally free in the noumenal world (the world of things as they really are, which I can never know by means of the senses or observation), whereas I am wholly determined in the phenomenal world (the world of the senses and of physical science).

Kant saw that if we accept a wholly deterministic and naturalistic view of the cosmos, there is a problem with asserting the existence of anything that seems to break those deterministic laws. That will include both human freewill and the existence of a God who acts in particular ways in the world.

Many philosophers are, like Kant, 'compatibilists' – they think that we must believe both in physical determinism and in freewill. They often do this by pointing out that I am free when I do what I want and am not compelled to do it. So I may be physically determined to want certain things, and to act in accordance with those desires, but I am still free if I can do what I want.

Kant was apparently not satisfied with this solution. He thought that, if I am to be morally responsible, I must be able either to do what is right or not to do it. That is, there must be alternatives to choose between. Nothing must determine my choice between those alternative paths except my own decision. Of course my decision can be *influenced* by many factors – by my abilities, my past actions and so on. But I am still free to do something not *determined* by any past event or law of nature. We can call that 'radical freedom', action undetermined by any prior physical cause. That is why Kant's position is particularly paradoxical. He wants people to be physically determined and radically free at the same time. This very odd theory led to Kant thinking that people were responsible even for the places and circumstances in which they were born. Once born, they were determined by the laws governing their empirical destinies. But in some mysterious way they had freely chosen the circumstances of their birth. Few compatibilists would agree with him about that – it is not really my fault that I was born in England – though there is perhaps no logical reason why they should not do so.

4. NEWTON AND DETERMINISM

I suppose most people would think it is impossible to be both physically determined and radically free. Either every event is physically determined or it is not. It is extremely difficult to decide this issue. But the problem partly arises because it is thought that Newtonian science requires a deterministic and naturalistic view of the cosmos (one that excludes any supernatural or spiritual causes). It should be realised, however, that for Newton himself it did not.

Notoriously, and wrongly, he believed that God would have to act in the universe at rare intervals to keep the planets in stable orbits around the sun. Otherwise they would after a very long period of time fall out of their orbits into the sun. It was this hypothesis that the French physicist Laplace had in mind when he is alleged to have said to Napoleon, when Napoleon asked him about God, 'I have no need of that hypothesis.' Laplace had refined the Newtonian calculus so that more precisely formulated equations of motion could dispense with the need for divine intervention. Thereby he got rid of what the Cambridge mathematician A.C. Coulson was to call 'the God of the gaps', a God who was needed to adjust scientific laws occasionally to keep the universe going (C.A. Coulson, *Science and Christian Belief*, Fontana, London, 1958, p. 41). It was partly Laplace's success in showing that such a God was not needed that gave impetus to the hypothesis of physical determinism. Laplace himself thought that everything that happens in the universe can be derived by sufficiently well-formulated laws from prior physical states.

This hypothesis still leaves a place for God as the author of the laws of nature, even though God is no longer allowed to act within the universe in ways that 'break' the laws. It is important to realise that Newton never accepted that the laws of nature are absolute, universal and unbreakable in this way. It is true that we can devise equations that describe the motions of physical bodies relative to one another. We can say, to take a simple example, that bodies attract one another with a force proportional to the product of their masses and inversely proportional to the square of the distance between them. But in what sense does such a law exist, even before any physical objects exist? Or what makes this law applicable to all objects in physical space, without exception?

Newton had an answer to these questions. He thought of laws as existing in the mind of God even before creation. God could compel planets to conform to such laws (or not, as God chose). But if God is removed, the laws of nature (the sense in which they can be said to exist, even before the existence of this cosmos) and the apparently necessary conformity of physical objects to those laws seem to become just highly improbable and inexplicable oddities of nature.

Perhaps the existence of laws of nature depends, after all, on the existence of God. If that is so, it is hardly possible to exclude just by definition God's action in the universe, miraculous or not. It is not surprising that for Newton himself the thesis of universal physical determinism was not acceptable. The laws of nature were not absolute and inflexible. They worked only by the will of God, who could abrogate them for good reason. Even if humans are not free in the radical sense of being sometimes not physically determined, God is surely radically free. God is never physically determined, since God determines all physical things to be what they are.

5. NEWTON ON THE NON-PHYSICAL BASIS OF REALITY

Far from being a believer in a wholly naturalistic and deterministic universe, Newton was convinced that science was not competent to determine questions about the ultimate nature of reality. He was content to describe the general laws that governed the observable behaviour of physical objects, at least in general and other things being equal. But he did not pretend to understand the ultimate nature of such laws, or of the reality upon which they operated.

Newton even found it very difficult to comprehend the force of gravity, although he had virtually discovered it. He wrote, 'That one body may act upon another at a distance through a vacuum without the mediation of anything else ... is to me so great an absurdity that I believe no man, who has in philosophical matters a competent faculty for thinking, can ever fall into.' Yet that is what gravity seems to be. Newton simply said that he 'framed no hypotheses' on this matter (*Principia Mathematica* [1686], trans. I.B. Cohen and Anne Whitman, University of California Press, Los Angeles, 1999, p. 943) and issued his law of gravity without pretending to know how it worked. What he himself had in mind was that God was the mediator who could accomplish action at a physical distance, but that was no part of his scientific hypothesis.

In modern physics, non-local action (action at a distance without physical connection) is a well-accepted part of quantum theory. We are perhaps still mystified about how it works, but there is little doubt that it does. It is noteworthy that Newton, virtually the inventor of 'laws of nature', did not give a materialist explanation for them, since he thought that such ultimate explanations were no part of 'experimental philosophy'.

He in fact thought that space and time were the 'sensoria' – the sensory apparatus – of God, by means of which God is immediately aware of objects. We humans have finite sensoria by means of which we are aware of objects in part and from a specific perspective. Newton believed that God had immediate knowledge of everything in the universe. The cosmos was completely dependent upon a cosmic intelligence, and this intelligence caused things to happen in ways that were beyond the reach of the natural sciences. The laws of physics register the observable and measurable principles of physical interactions, but do not pronounce upon the ultimate nature of reality or upon the ultimate reasons for things being as they are.

When it came to speculations about ultimate reality, Newton undoubtedly believed in a spiritual basis to the physical universe. A good account of his views can be found in Richard Westfall's *Never at Rest: a Biography of Isaac Newton* (Cambridge University Press, Cambridge, 1980); and a beautifully produced book, *The Newtonian Moment*, by Mordechai Feingold (Oxford University Press, Oxford, 2004) provides fascinating information about his cultural context. Newton wrote at great length about Hermetic philosophy and alchemy, and a long and boring commentary on the prophecies of the Book of Daniel. I have to say that I would not recommend Newton's writings on these topics to any normal person, though they are of interest to historians. They are of no more than historical curiosity now, but they show how far Newton was from being a materialist.

The virtual inventor of the laws of physics saw them as sketching the mechanics of the universe insofar as it is purely physical, in abstraction from spiritual influences. But in fact he thought that the whole physical universe, though it is supremely beautiful and mathematically elegant, is only an appearance of a hidden underlying reality of a spiritual nature. Newton believed that, for those who have eyes to see, that hidden reality even makes itself known in the mysterious operation of gravity and the movements and disposition of the planets.

The laws of physics do not show nature as it really is; they show how nature operates, when treated in abstraction from spiritual reality. The laws of nature are the laws of God, a God of supreme rationality, intellectual beauty and intelligence. Newtonian science may have led to the belief that the universe is a machine without a mind, but that was not this science's origin or its motivating force. For Newton, the mechanisms of nature are but the partial appearances of that which is most real and vital, the operations of the mind of God.

3

THE LIFE OF THE UNIVERSE

Newton's rather static view of the universe has since been succeeded by the idea of a much more dynamic, developing and self-organising cosmos. This is a universe with a history, a perspective that encourages a greater interest in human creativity and hope for the future. It draws attention to the immense amount of fine tuning necessary for life to emerge. For believers in God, it suggests a view of the universe as a gradual and progressive unfolding of potencies always present in the universe but becoming at least partly self-directing in the course of time. It provides a model for 'creation' that would have been quite unthinkable before the rise of modern science, but is now of great importance for religious belief. For it seems that even God has a history, partly played out in the story of this cosmos.

AIM: To show the 'fine tuning' of the universe for life, suggesting a divine purpose for the universe. To show that humans are not 'slaves of nature', but that they have a responsibility for co-operating with God's purpose by shaping the natural world to encourage greater personal flourishing.

1. THE FINE TUNING OF THE UNIVERSE

Isaac Newton discovered that physical objects exert a force of attraction (gravitational force) throughout the universe, which is weak but infinite in extent. Since Newton's days, physicists have discovered further fundamental forces – electromagnetism and strong and weak nuclear forces – that bind all entities in the universe together in patterns of regular interaction.

These forces are universal and constant, and that constancy makes possible the formation of atomic structures from simpler sub-atomic particles. Such structures are remarkably stable and they are able to form the basis for further agglomeration into compound molecules of immense complexity. These can replicate and eventually assemble proteins to give rise to organic bodies. The propensity of relatively simple particles to assemble in layers of highly structured and interdependent complexes – bodies and central nervous systems – within which consciousness and understanding are generated is amazing. But it all seems to happen in accordance with the basic properties of the fundamental forces and particles of nature, which are mathematically simple and related in just the precise correlations needed to give rise to life and consciousness.

The simplicity, correlation and constancy of the fundamental forces of nature, giving rise to the complexity, structure and variety of material entities, and, it seems, to consciousness and intelligence, are so astounding that they strongly suggest design. We could always say that they just happen; they are just ultimate brute facts. But if a great mathematical genius wanted to set up a program for generating a large variety of complex and beautiful forms with as few elegant basic rules as possible, the laws of nature would fit the bill excellently.

At the end of the twentieth century, new discoveries in physics brought out for the first time just how very extraordinary the nature of our universe is. The fundamental forces and constants of nature need to be very closely attuned to one another in order to give rise to stable atomic structures, to stars and planets, to organic life and to consciousness. Much has been written about the 'fine tuning' of these fundamental forces, and I will mention only three examples of such fine-tuned forces. Fuller accounts can be found in many places. One of the most thorough accounts is: John Barrow and Frank Tipler, *The Anthropic Cosmological Principle* (Clarendon Press, Oxford, 1986). John Polkinghorne's *Reason and Reality* (SPCK, London, 1991) and other works provide shorter, readily readable expositions.

One example of fine tuning is that in the early universe, probably about fourteen billion years ago, the forces driving cosmic expansion and the force of gravity needed to be precisely balanced (balanced to one part in 10 to the power 60, an absolutely enormous number) if the universe was not either to expand too quickly to allow stars to form, or to collapse under its own gravitational pressure. It looks as though stars, galaxies and planets could not have formed at all if the forces of cosmic expansion and gravity were not tuned to one another with absolute precision. It could, of course, have happened by accident, but it is not just rather improbable; the improbability of this happening is vanishingly small!

In a similar fashion, the forces of gravity and electromagnetism must be precisely balanced if stars were not to burn themselves out before life had time to evolve, or burn so feebly that they would not produce enough energy for life to evolve.

Thirdly, the nuclear force within atoms needs to be in a very precise proportion to the electric force. If it were slightly weaker or stronger, elements like deuterium that are vital to the chain of nuclear reactions that keeps the stars burning would not form, and the heavy elements necessary for life would never come to exist. Again the evolution of life depends upon a precise balance between the fundamental forces of the cosmos.

2. THE MANY WORLDS INTERPRETATION

There is a long list of such instances of the fine tuning that is necessary if life is to evolve in the universe. Martin Rees lists some in *Our Cosmic Habitat* (chapter 11). Hugh Ross has listed twenty-five parameters for the universe that must have values falling within very narrowly defined ranges for life of any kind to exist (Hugh Ross, *The Creator and the Cosmos*, Navpress, Colorado Springs, 1993, pp. 111–113). One conclusion that might be drawn is that life depends upon a combination of so many improbable factors that the whole cosmic process must have been set up by some vast cosmic intelligence. This is a natural enough conclusion, but the data do not force anyone to conclude that there must be a cosmic mind that does the fine tuning.

One widely canvassed alternative explanation, suggested by Hugh Everett in 1957, is that every possible universe actually exists. This universe is part of a much larger collection of universes, a 'multiverse'. The multiverse could consist of every possible universe there could ever be, all of which actually exist. If this were true, then, however incredible this universe is, it is bound to exist, and no recourse to a cosmic mind is needed.

But the idea that every possible universe is actual may well seem more extreme than the postulate that there is a cosmic mind. Moreover, we would still have to ask what is meant by 'every possible universe'. Is there supposed to be a set of all possible universes? Is that set finite or infinite? What is it that determines the nature of these universes? Is there some power that can ensure that they all get actualised? And where is that power supposed to exist?

If there is such a set, one obvious place to put it is precisely in the mind of a cosmic creator who envisages all possibilities and has the power to actualise them. In that case, however, they will almost certainly not all be actualised, since a creator will be able to choose which ones to actualise. If the creator has any decency, the possibly enormous set of extremely bad universes, in which there is endless and totally pointless suffering, will not be actualised. Some selection between possible universes will be needed.

Martin Rees, while issuing a 'health warning' against heady speculation, finds some force in the suggestion that there exists a set of possibilities, and that in some sense they 'press into existence', as Leibniz put it. I think this mainly appeals to Professor Rees because it enables us to explain why a fine-tuned universe exists without appeal to God – 'if our universe is selected from a multiverse, its seemingly designed or fine-tuned features would not be surprising' (*Our Cosmic Habitat*, p. 165). But we still have the major problem of the sense in which such possible worlds really exist, and of how some (or all?) of them get 'selected' for actuality.

Possible worlds could exist in the mind of God, and God, the cosmic mind, could select some for actualisation. This universe may indeed be one among many that are 'chosen', but if the creator is reasonably good, it will be such that the specific forms of intelligent life that exist in it (us) exhibit some distinctive sort of goodness. What the fine-tuning argument seems to show is that we, or forms of life that are closely similar to us, could only exist in a universe with the fundamental constants and forces of this universe. So if God selects us for existence, God has to select a universe with just the sort of fine-tuned parameters this one has.

3. THE PLACE OF PERSONS IN THE UNIVERSE

These developments in post-Newtonian physics justify two major and surprising claims. First, this universe is fine tuned for intelligent life with amazingly improbable precision, and this is at least compatible with and even strongly suggestive of cosmic design. Second, the forms of intelligent life that we know as human persons could only come to exist in the framework of this universe, with its evolving history and general structures of law. We are products of a long process of cosmic evolution, in a universe that is precisely tuned for our existence. We could not have existed as the precise persons we are in a supposedly better universe. We are formed of the dust of long dead stars, and of atoms and molecules that can only exist in a universe with these virtually unique parameters. If our existence depends upon the fine tuning of this cosmos, then if that fine tuning did not exist we could not exist either.

An example of the way in which our existence depends upon some rather surprising features of the universe is the fact that the universe has to be the size it is in order to have us in it. Some people have been depressed by the vast amount of empty space there seems to be – our expanding universe extends for billions of light years. That seems to make human life very small and insignificant. But though it seems very odd at first, it turns out to be true that without as much apparently empty space as there is in the universe we could not exist at all. For the extent of space is the result of the expansion of the universe, which is in turn the result of the inflationary and gravitational forces at work in the early universe, which is a condition of our coming into existence. We should not get depressed by the vast amount of empty space there is. It is there because that is what it takes to enable us to exist at all. It looks as if we have to exist in just this universe, whether we like it or not. Even if we may subsequently exist in a better world (in a Paradisal world, for example), we have to have been born and to have grown in this universe for us to be the beings we are. Either we are parts of this universe and its fundamental forces, or we could not exist at all.

These claims suggest that this universe exists both for the sake of the elegance and beauty it manifests and also for the sake of the specific forms of intelligent life it produces, which can actualise new sorts of consciousness and creativity that would not otherwise have existed. Putting Newtonian laws in the more modern perspective of cosmic evolution, we can gain deeper insight into the place of humanity in the cosmos and into how the basic forces and laws of the universe necessarily have the values they do in order to produce a universe like this, with beings like us in it. Those laws generate intelligent agents, parts of the natural order, that are conscious of and able to appreciate the beauty and intelligibility of nature, and that can create new forms of beauty themselves. We might say that the cosmos comes to understand itself and to shape itself in new ways. Nature comes to consciousness and responsible self-direction, and so discloses that such things have always been potential in the natural order itself. It looks as though the history of the cosmos is the story of a development from a primitive state of extreme simplicity and unconsciousness to a state in which highly organised complexity, consciousness and self-direction exist.

This could just be the way the universe is, without any reference to a creator God. It is a view nobody could have taken seriously before the rise of post-Newtonian physics. But it is compatible with the existence of a creator, and to some of us it positively suggests that there is a creator. Isaac Newton thought the universe was the handiwork of God, a great spiritual intelligence. It does seem possible that such a God could choose to create a cosmos that, through long development and partial self-organisation, realises its own potential by a sort of self-shaping. As the novelist and Christian priest Charles Kingsley (author of *The Water-Babies*) said, 'God makes things to make themselves.' Societies of intelligent beings generated within the physical universe can come to understand the hidden nature of the cosmos and of its creator. They could perhaps come to mediate the power and wisdom of the creator in their own lives, gradually learning to use the laws of the natural order to further the realisation and flourishing of personal lives.

4. NEWTON'S GOD

Newton did not have this more modern perspective of cosmic evolution. But he did think the universe was the creation of a super-intelligence before he ever set out on his scientific investigations. That belief motivated him to seek the simple and reliable laws upon which such a cosmos would be constructed – and he found them. Newton is sometimes regarded as the inventor of the Great Disembodied Mathematician, who sets up the universe and then ignores it. This is the Newtonian God portrayed by Blake in his engraving *The Ancient of Days*, and it is 'the God of the philosophers and men of science' disparaged by Pascal in his 'Memorial'.

In fact, Newton's God was a personal God who desired the salvation of all humans, who wished that they would come to know and love God and who set up the universe to make that possible. Of course the whole cosmos is not just a means to this end. As we have seen, it has its own beauty, and it would be reasonable to think that its beauty and intelligibility are of value to God for their own sake. But the cosmos achieves its most complex and integrated form in communities of sentient beings of intelligence and freedom. One goal of creation, Newton thought, is that in such communities, knowledge and love of the universe and of its creator should come to exist.

Newton's own view of the universe, therefore, was not that it is deterministic and naturalistic. That is an exaggeration and misrepresentation of his scientific claims. What Newton's position requires is that there are constant and reliable laws in accordance with which nature operates, insofar as it is not affected by personal factors, such as radical human freedom or special and unpredictable divine action. The Kantian split of the world into two parts, one subject to scientific study and the other concerned with postulates of morality and religion, is not necessary. It is the same world that operates according to law and is directed in accordance with purposes, both divine and human.

For Newton, there is no problem of divine action in the world. God can act by direct effective volition at any time. God made the laws of nature, but is not bound by them. The only problem of divine action is why it is that God does not act by direct volition more often. The Newtonian and post-Newtonian view of a law-governed, self-organising, developing cosmos helps to resolve this problem. God will set the cosmos on its way towards its divinely intended goals, but the cosmos operates in accordance with its own law-like and intelligible principles of development (which are of course themselves created by God). Only thus can finite persons come to understand it and partly control it, and take responsibility for their own personal growth. Only thus can the cosmos be, or give rise to, a form of being that is truly independent and autonomous.

We might expect that, all the same, God would act to ensure the realisation of the divine purpose, and to make it possible that all finite persons might come to know and share in the love of God. Whether this means that God has to perform some spectacular miraculous acts, as Newton believed, we do not know. But it may well mean that God is somehow actively involved in the causal processes of the universe to ensure that it reaches its appointed goal.

5. COSMIC PURPOSE

If the creator has set up the cosmos so that it, or parts of it, can achieve personal union with God, then its basic laws will from the first be oriented to that end. The cosmos will not be a closed causal system. Its laws will be essentially open to personal influence and they will be oriented beyond themselves to a reality that is not in space-time. The difficulty with the machine analogy, generated by some followers of Newton but not by Newton himself, is that it seems to make the universe lack intrinsic value, being merely instrumental to some purpose beyond it – as a machine is instrumental to some purpose external to the machine itself. But the universe has value in itself, as an object of the divine consciousness and a manifestation of divine wisdom, beauty and power.

If the universe has a purpose, it is internal to the cosmic process, lying in the realisation of values it generates. This makes it more like a work of art than it is like a machine. Perhaps the closest artistic analogy, suggested by the physical biochemist and theologian Arthur Peacocke, is that of improvised music, in which players listen to others and add their own creative contributions to the work, though all follow the general directions and outline pattern set by the composer. (See Arthur Peacocke, *Paths from Science towards God*, Oneworld, Oxford, 2001, p. 77. See also Arthur Peacocke and Ann Pederson, *The Music of Creation*, Fortress, Philadelphia, 2005).

The performance of such music requires that sounds can be predictably and reliably produced and that there are stable laws governing the production and reception of sounds. But it would be absurd to say it requires that all music is determined, that it could not be other than it is or that no non-physical, personal, factors can be involved in a performance.

6. NATURE AND COSMIC EVOLUTION

In some traditional Christian views, where God creates the cosmos solely for the sake of humans, there would seem to be no reason why God should not just create humans immediately, without going through the long process of cosmic evolution. But we can now see that if the cosmos gradually realises its potentialities through a process of gradual and continuous development, that may have a quite distinctive value of its own. Humans, and perhaps other intelligent agents too, are generated as part of this cosmic system, and so they are in general subject to its laws of operation. To understand human nature is to understand the fundamental laws of the cosmos as a whole, and how they operate to generate carbon-based life forms of a specific anatomical structure. We cannot see the cosmos as simply subordinate to human nature. We have to see humans as minute but highly developed parts of a cosmos that has its own significance.

Newtonian science reinforces the Galilean reversal of medieval Christian cosmic priorities. The universe does not exist for the sake of humans. But humans might have a specific role and responsibility in the universe. They may play a part in implementing the purposes of the creator and actualising them in new ways. They may know and appreciate the values the universe actualises, and that, for a religious believer, will be an important part of the worship of God.

This is not quite a 're-enchantment' of nature, since nature is no longer seen as sacred, inviolable and perfectly in order as it is. Nature is rather seen as a beautiful, elegant and deeply intelligible system, of objective value because there is a divine consciousness that values and appreciates it for its own sake; yet Nature is not by any means perfect. Nature is a developing system, and the very generality and elegance of its laws bring suffering and pain to many sentient beings. So in the end the processes of nature are to be subjected where possible to the flourishing of conscious beings, and it is a human responsibility to subject natural processes to the requirements of the well-being of sentient creatures, through learning, struggle, training and discipline. Nature is not sacred, but the religious believer will treat it with respect and care, because it is chosen and created by God.

The Newtonian worldview is not, as is sometimes said, that the universe is a deterministic machine that, once set up, proceeds on its way without further reference to its creator. The universe is an elegant, intelligible and reliable structure that generates increasing levels of organised complexity, increasingly open to personal knowledge and action. It is a realm of values to which humans can contribute their own unique forms of appreciation and creative action. The laws of nature set out the basic reliable principles upon which humans can depend to enable them to do so.

The emphasis of Newtonian science is not on the reduction of humans to parts of machines. It is on human freedom, creativity and responsibility. Yet the consequences of the Newtonian revolution have split in two quite different directions. On the one hand, reductionist science sees the universe as a deterministic machine, in which every event can be explained, in principle, by appeal to simple atomic particles and a few basic absolute and unbreakable laws of nature. Newton himself would definitely have rejected this view, but the view can still be found among some philosophically minded scientists.

On the other hand, the universe can be seen as an intellectually beautiful and intelligible composition of a supreme intelligence. It is not either wholly deterministic or mechanistic, though it normally operates in accordance with predictable laws that are well fitted to human understanding. For some Platonically influenced medieval views, the highest human wisdom lay in turning away from nature to inner contemplation. After Newton we are encouraged to turn back to nature, to admire and appreciate the universe as God's handiwork. This is a definite turn towards the world as a proper object of investigation and contemplation, and at least for Newton it is part of the proper worship of God.

In early medieval times, humans just had to accept with resignation the way things were, since it was believed that God had made the universe and God's work should not be interfered with. But for Newton humans can modify nature. The service of God lies not in accepting with resignation the way things are, but in remaking the natural world to minister to the needs of sentient creatures. This is a much more active and interventionist view of human life in relation to nature.

Newton's revolution was not only a breakthrough in scientific understanding. It changed the whole human approach to nature, and to human responsibility for the future. Virtually for the first time in history, humans could change the world for the better. For the reductionist point of view, this meant that humans were free to do anything they wanted with nature, to use the natural world for their own pleasure – something that has turned out to be a dangerous prescription for the world and for the human future. But from Newton's own religious point of view, it meant that true religion lies not only in turning from the things of time to contemplation of the timeless and unchanging. It lies in co-operation with the creative power of God as it continually realises new values in history. Humans have power over the future, but they are to exercise that power with care for the world and responsibility to God.

7. A SCIENTIFIC SPIRITUALITY

The turning towards history and time, the stress on human creativity and responsibility for the future, and the hope that the world can be changed for the better are signs of a new attitude to spirituality for which the birth of natural science is largely responsible. No longer is the realm of faith seen as the realm of the timeless, the resigned, the otherworldly. Now it is the realm of the temporal, the creative and the world-transforming. And is all this because of Newtonian mechanics? Not quite, but both Newtonian mechanics and the new attitude to spirituality are expressions of a new approach to nature and to human existence.

Both give rise to the temptation to go further, and say that this world is all there is, we must freely invent our own meanings and purposes and we can make the world as anything we please. The Faustian alternative to faith is born – like Dr Faustus, we claim the power to do anything we like with the world of which we are part. But that Faustian alternative should not be called part of a scientific worldview, whose basis is a profound faith in the human capacity to understand an intelligible universe, and a belief that it is possible to adapt nature to purposes that are of intrinsic value and are rooted in human nature.

It is one of the paradoxes of the scientific Enlightenment that it gave birth both to an enhanced belief in human autonomy and power over nature and also to the belief that humans are just parts of a universal deterministic machine – and therefore powerless to do anything other than whatever the blind and purposeless laws of nature decree. Newton's own vision does not split the world into two incompatible parts in this way. For him, the laws of nature in their beauty and intelligibility manifest their source in the personal, mind-like reality of God. Values exist objectively, as states that are appreciated as intrinsically desirable by God. And those values lay down the goals and limits of human action in the world. Yet humans have a creative part to play in the actualisation of many new values, and scientific spirituality lies in the combination of awe before the wonderful intricacy of nature and the acceptance of responsibility for using the laws of nature to actualise new and distinctive forms of value.

8. THE HISTORY OF THE COSMOS

Newtonian mechanics, it is often said, takes no notice of temporal flow. That is true, in that the equations of motion can run either forwards or backwards indifferently. But as astronomy and cosmology have progressed since Newton's day we have become much more aware of the way in which the cosmos has developed and parts of it at least have become much more complex over time. Whatever difficulties there are in quantum theories of the beginning of the universe, it seems firmly established in modern cosmology that 'at about 15 thousand million years ago the material content of the universe was compressed into a tiny region from which it expanded at great speed against the force of gravity' (Chris Isham, 'Quantum Theories of the Creation of the Universe', in *Quantum Cosmology and the Laws of Nature*, ed. Robert Russell, Nancey Murphy and C.J. Isham, Vatican Observatory Foundation, Vatican City, 1993, p. 50). The classical big bang theory pictures the universe as expanding from an initial singularity of infinite density and temperature. There are currently a number of different cosmological theories that seek to explain this ultimate 'beginning' of our space-time. But all agree that our universe expanded from a very early state (the 'tiny region') in which there were no complex physical entities at all, to its present condition, including the extremely complex self-replicating molecules and neuronal circuits of the human brain. The universe has a history. It was born and it will die. It achieves a maturity in which many complex possibilities are actualised, and then, in accordance with the laws of thermodynamics, it begins to decay, until at last, after billions of years, it ceases to be. Whether it ceases in a big crunch or a long drawn-out heat death, whether or not it is succeeded by a series of 'bouncing' – expanding and contracting – universes, whether or not it somehow generates out of black holes other forms of space-time, as Lee Smolin imagines (in *The Life of the Cosmos*, Weidenfeld & Nicholson, London, 1997), this space-time regime in which we exist will certainly die.

If there is a God, then that whole history is forever present to the mind of God. It is not forgotten or set aside, and so it possesses a sort of eternity. Yet such an eternal reality in the mind of God would not have existed had the universe not existed. There might be possibilities of infinitely many universes in the divine mind, and God may find satisfaction in contemplating them. But they will not be actual; they will be things that could happen, not things that are real. Any conscious beings there might be in those possible universes will not actually have any thoughts or feelings. So what God knows and experiences if there is an actual universe differs from what God knows if a universe remains only possible.

In that sense the existence of the universe makes a difference to God. The universe becomes part of what God is, for God's knowledge is certainly part of God's being. And so we might say that the history of the universe is part of the history of God. It is one way, perhaps one of indefinitely many ways, in which the being of God is manifested. Out of all the possibilities inherent in the divine being, all the possible universes that could ever be, this universe is actualised (perhaps, though we do not know, one of indefinitely many). So it can be seen as part of the self-realisation of God. It is God actualising some of the potencies in the divine being, recognising and appreciating that actualisation, and taking it back into the divine being as a completed and perfected experience.

The development of science in the seventeenth and eighteenth centuries left two very different inheritances. One was the reductionist, determinist worldview in which any idea of God disappears. The other is a revised view of God. The medieval view was that God is perfect, changeless, the ultimate cause of all that happens, unaffected by anything that happens in the world. But after the rise of the new scientific emphasis on the importance of cosmic development and temporality, it becomes possible to see God as a being with a history, who can be affected by the universe and who gives finite persons a role in creating the future. Either way, science does not leave religion untouched. Either God is dead, or God gets much more involved with time and change. The issue is not yet resolved. But there are further scientific revolutions to come.

The sense of the cosmos as an unfolding of potencies implicit in the divine being gives a sense of the universe, not as a static, complete actuality, but as a dynamic, developing and at each stage partly incomplete realm of becoming. It is this sense that underlies one very significant part of the scientific worldview, the theory of evolution.

4

THE EVOLUTION OF
LIFE ON EARTH

*Perhaps the biggest change that modern science calls for in trad-
itional religion is the acceptance of evolution, both cosmic and
terrestrial. Some biologists represent evolution as a combination of
chance and the blind sifting processes of natural selection. But it
can with equal plausibility be seen as a supremely elegant process
directed to goals of intrinsic value. God sets the goals and designs
the process for achieving them, which works out over aeons of
time. Religious views must now be set within this wide evolutionary context.*

*AIM: To show that the evolution of organic life forms is compat-
ible with religious belief. To show that, though biologists do not
expressly deal with the notion of purpose, it is nevertheless reasonable to see a goal in evolution.*

1. TWO DARWINS

The third great revolution in human thinking brought about by science is the establishing of the theory of evolution. Most people associate the theory of evolution, quite rightly, with the publication, in 1859, of Charles Darwin's *Origin of Species*. But evolutionary theory, in a less scientifically evidenced form, had existed for some time before that. Charles's grandfather, Erasmus Darwin, was an evolutionist. In 1801 he wrote, 'All nature exists in a state of perpetual improvement ... the world may still be said to be in its infancy, and continue to improve for ever and ever' (*Zoonomia*, A. Johnson, London, 1801, vol. 2, p. 318). Erasmus saw in this the wisdom of the creator: 'That there exists a superior Ens Entium, which formed these wonderful creatures, is a mathematical demonstration. That he influences things by a particular providence, is not so evident' (letter to Thomas Okes, November 23 1754, in Michael Ruse, *Monad to Man*, Harvard University Press, Cambridge, MA, 1996, p. 62). Particular providential interventions were, he thought, not necessary, since evolution could proceed simply by the operation of general laws.

The idea of perpetual improvement is one that finds echoes in his grandson Charles ('All corporeal and mental endowments will tend to progress towards perfection': *Origin of Species*, Penguin, Harmondsworth, 1985, p. 459), but in him it is tempered by more cautionary thoughts that maybe the course of evolution does not run smoothly from good to better for ever. The basis of evolution, in Charles Darwin's theory, is 'descent with modification' plus natural selection. Though he had no idea of the mechanisms of heredity, Charles observed that animals tended to have large numbers of offspring, all of which differed slightly from their parents, some in quite noticeable ways. In situations of scarcity, there was a competition for survival, and some variations were more useful to survival than others. Thus over many generations there would be a selection by the environment of the better-adapted varying forms, and that was why different environments, for instance in the diverse islands of Galapagos, produced rather different adaptations of what had been the same species.

2. CONTINUAL IMPROVEMENT?

In 1953, James Watson and Francis Crick discovered the now famous double helix structure of DNA, the chemical carrier of heredity. From that date the new sciences of molecular biology and molecular genetics began to provide a detailed chemical explanation of Darwin's rather vague postulate of 'descent with modification'. Now we can say with much greater confidence that evolution proceeds by a combination of random mutation (often called by biologists 'mistakes' made in copying strands of DNA) and natural selection. The fundamental mechanisms accepted by modern biologists for evolutionary change remain those of continual, but usually slight, variation in the structure of the DNA, which instructs proteins to build organic bodies, and the elimination of less well-adapted organisms or the selection of better-adapted organisms by environmental factors. Whether this leads to continual improvement or not is another question. There does not seem to be any particular reason why it should do so, but it is generally felt by biologists that the competitive pressures for scarce resources will continue to select the better-adapted members of species, and over a long period of time organisms will adapt well to their environment; in this sense – but perhaps only in this very restricted sense – species will 'improve'.

We can intuitively see how better mobility, keener senses and greater intelligence might well be successful adaptive factors. They might enable a species to slaughter, evade or out-think other species in the search for food much more effectively. The way some rather depressive biologists see it is that in the 'arms race' between species, the ruthless life-or-death competition that characterises life on earth, the fittest will win in the end, and in the desperate struggle for life that drives evolution we might expect improvement of a sort – though whether it would be a very admirable improvement is quite a different issue.

3. NEGATIVE INTERPRETATIONS OF EVOLUTION

This Darwinian theory of evolution looks very different from the theory that the universe is a manifestation of possibilities inherent in a divine and perfect being, a manifestation that develops towards ever-fuller actualisation of new and creative values.

A negative take on the scientific worldview would see the demotion of humans from a central place in the universe as the loss of a sense of the significance of human life. Newtonian mechanics had already seemed to some to reduce the universe to a mechanical circulation of minute parts, grinding away without point or purpose towards an inevitable dissolution. As Steven Weinberg has put it, 'the more the universe seems comprehensible, the more it also seem pointless' (S. Weinberg, *The First Three Minutes*, Andre Deutsch, 1977, p. 149). And now Darwinism completes the demolition of optimism by pointing out that humans are no more than the accidental results of millions of mistakes in the copying of DNA, that they only exist because they have exterminated all competitors on the planet, and that there is no reason why they should continue to survive for much longer, since their excessive aggressiveness will in all probability lead to their imminent self-destruction.

This is depressing indeed, but it is a good example of how evaluations get imported into scientific findings and spoken of as though they were facts, whereas the facts allow a very different evaluation. The facts of evolution, which are now beyond reasonable scientific dispute, are that all present life-forms on earth have developed from simpler organisms over a period of about four billion years, and that the main mechanisms of this development have been changes (mutations) in molecules of DNA, plus selective influences in the environment that have encouraged the emergence of specific types of organism.

4. EVOLUTION AND MORAL COMMITMENT

Many traditional religious views, including the Christian, have held that there was an age of innocence, without suffering or death, that degraded into the present more corrupt state of affairs on earth. Evolutionary theory reverses this judgement. Suffering and death were part of organic life for millions of years before humans existed. Without suffering, the capacity for feeling could not exist, and without death the development of new species would not be possible. There was no golden age that we have lost. Our ideas and beliefs have developed from primitive and non-rational beginnings towards greater understanding and rationality (sometimes). Humans have not grown in moral sensitivity as much as they have in technical skill – though even in the area of morality there have been some advances in some societies, for instance in attitudes to women, slaves and animals.

What this suggests is that intellectual growth is different from moral growth. The former can be used in the service of personal or kin-group superiority over others, in obtaining the satisfaction of desires more efficiently, at whatever cost to others. The latter is more a matter of extending natural feelings of compassion to include groups very different from your own. There is no compelling reason to do this, except that anyone can see that it is, abstractly speaking, the most fair or just thing to do. But why should that influence a purely rational agent in a radically unjust world? Is it not more rational to seek to satisfy your desires as well as you can, and make sure that others will not have the power to prevent you from doing so?

Our view of the nature of the cosmos, and of the human role within it, might well affect our response to this question. If we had an optimistic view of evolution, we might think that the cosmos is founded on a nisus towards the flourishing of all creatures, which it is a human responsibility to implement, and which is destined to be actualised at some time. There is a cosmic goal, there is a human role in relation to it in which the meaning of human life lies, and there is a positive hope for its realisation. With these beliefs, moral development might have a chance of keeping pace with intellectual development.

Unfortunately, the rise of science seems to many to have undermined belief in a cosmic goal (the universe has no purpose discoverable by science), in any objective significance for human life (we are accidents of evolution) and in any ultimate cosmic hope (all things will end in the entropic death of this cosmos).

If humans are accidental products of a purposeless process, it seems strangely irrational to suppose that they have a quite unique dignity and moral status. It seems odd to say that there are absolute and objective moral obligations, when morality is so obviously an invention of a bunch of primates crawling around on the surface of a small and insignificant planet. There may be many reasons why belief in moral obligation exists, and evolutionary biologists may give a plausible account of the development of moral beliefs. I shall be talking about such accounts later (in chapter 12). But once the evolutionary mechanisms giving rise to moral beliefs have been exposed, the question becomes: is it rational to follow those beliefs as though they were absolute moral obligations, when I know they are not?

Once you know the mechanism, it becomes irrational to simply allow it to dominate life. Perhaps you can change it, or break free of the mechanism. It is at that point that it becomes possible to liberate oneself from ancient moral taboos and choose a rational course of life with fully responsible freedom. Since all who know this will choose freely how to live, it is difficult to predict what might happen if everyone accepted such a view. But, speaking for myself, I would think it most reasonable to try to maximise your own desires, while trying to retain the admiration of those few people you care about. I also think that this is how most humans have acted for most of their history. That is precisely why the world is in such a catastrophic mess.

One of the things religion has done is to argue that there is a rational and objective basis for seeing humans as being objects of unique moral respect and for thinking that there are absolute moral obligations. Admittedly, the testimony of religion has been compromised time and again by the hypocrisy of its leaders and the alliances it has made with political power, by repeated failure to follow its own most basic precepts.

Yet, taking what I have called the pessimistic view of evolution, it can seem that now science has undermined all religious arguments for the objectivity of morality, and shown moral beliefs to be the result of successful early evolutionary strategies that are now often counter-productive. The way is open to the naked pursuit of power, the triumph of desire and the supremacy of the will.

It is paradoxical that science may thus seem to challenge morality at the very point at which science advocates selfless devotion to the pursuit of truth, suppression of any desires that may replace unwelcome truths by easy wish-fulfilments, and trust both in the community of scientific colleagues and in the rationality of the universe itself. Science can be highly morally demanding. At its best, it expresses an absolute commitment to truth and understanding, even when it seems to threaten the traditional foundations of morality.

This suggests, I think, that science should be seen as a strong support for an objective morality of truth-seeking, ruthless honesty and trust in the intelligibility and beauty of being. Something must be wrong with a scientific account of evolution that threatens the very existence of such an objective morality. It is one of the great paradoxes of modern scientific thought that an absolute commitment to truth should end by suggesting that truth is not really worth pursuing unless you happen to feel like it.

The resolution of this paradox lies in the perception that the undermining of any justification for a truly binding morality does not belong to the practice of science itself. It belongs to the realm of pessimistic evaluations of scientific theories, to a form of reductionism that, if pursued relentlessly, would undermine the scientific enterprise itself. Such evaluations are in fact foreign to the scientific enterprise, as is shown by the fact that precisely those who loudly proclaim the pointlessness of the universe have often spent their lives in a selfless devotion to truth and felt awe and amazement before the beauty and intelligibility of being.

5. A POSITIVE INTERPRETATION OF EVOLUTION

The process of evolution, as we now see it, is a result of the operation of physical laws of great mathematical beauty and intelligibility. It follows that there can be no real accidents in evolution. If the universe is beautiful and intelligible, and its contemplation by consciousness is of value in itself, is it an accident that it has given rise to forms of finite consciousness that can, by effort, striving and discipline, come to understand and appreciate that beauty and intelligibility? Even from a strictly scientific point of view, we can see the universe as an elegantly structured process that has generated out of itself beings of consciousness and freedom. Since value consists in the appreciation of sensory and intellectual beauty, these beings actualise new forms of value, or are capable of doing so. The process of cosmic evolution is thus one that proceeds by elegant stages from an origin of supreme simplicity (the big bang) in which there are no actualised values (except perhaps to God, as contemplator of the cosmos), to complex highly organised states of many sorts of interconnected, intrinsic and actualised value.

Perhaps a scientist would be cautious about saying that any cosmic process has a purpose. After all, there is no observable agent who is acting to realise a desired goal. Nevertheless, if a purposive process is one that moves by efficient steps to a valued goal, the process of cosmic evolution undoubtedly *looks* purposive. So far, perhaps surprisingly, this argument is in agreement with the high priest of modern atheism, Richard Dawkins, who writes, 'The living results of natural selection overwhelmingly impress us with the appearance of design as if by a master watchmaker, impress us with the illusion of design and planning' (*The Blind Watchmaker*, Penguin, Harmondsworth, 1986, p. 21).

Dawkins stresses that this is only the appearance or illusion of design. For he has a story to tell of how it can all be explained by random mutation and non-random but nevertheless 'blind' natural selection, without any appeal to a designing mind. Even if this is true, it does not entail, however, that the appearance of design is illusory. If it looks awe-inspiringly designed, then it may be so. The fact that the design is achieved through mutation and selection does not show there is no real design. That is what the design might be!

The appearance of design does not prove there is a designer. But it does suggest that this universe is the sort of universe that could be designed by some sort of wise intelligence. This is the sort of process that could be set up to realise the highly desirable goal of communities of finite beings understanding and delighting in the nature of the cosmos of which they are part. The cosmos moves towards its own self-understanding, and that is not pointless. It is precisely the point, the value and meaning of being as it exists in this physical universe.

6. IS EVOLUTION RANDOM?

How, then, can an evolutionist say that the process is just an accident, or that it is purposeless? Partly this is just a blunt assertion that no one planned or envisaged it, that there is no God or ultimate cosmic mind. Darwin thought that the cosmos certainly suggests such a mind. He just came to doubt the benevolence of that mind. The more that improbable accidents happen to someone, the more we begin to suspect that there is more than pure chance at work. And there are just too many improbable 'accidents' in a universe that produces intelligent, feeling beings out of clouds of interstellar dust – millions of fortuitous accidents – for us to be sure that it is all pure chance.

Richard Dawkins is adamant that he is not saying it is all pure chance. Genetic mutation is random, in the sense that 'mutation is not systematically biased in the direction of adaptive improvement' (*The Blind Watchmaker*, p. 312); but natural selection is non-random, in that it selects, by a process repeated time and again over a very long period, organisms that are more reproductively successful. This cumulative selection has no long-term goal, but it will progressively select organisms that come to have the appearance of design.

I have no objections to this illuminating account of evolution. But I would observe how remarkable it is that random mutations should continue over millennia to generate some organisms that are capable of becoming cumulatively more complex. And how remarkable it is that the environment should be precisely such as to select accumulated 'improvements' that eventually lead to the development of brains and consciousness and intelligent moral agents. This is not to say that God is required to account for evolution. But it is to say that, as in the case of the fundamental constants and forces of nature, so the evolution of organic life requires a great deal of fine tuning in the sorts of mutation that occur and the sorts of environment that exist, in order to produce intelligent life. There is an exceedingly small window of opportunity within which mutation and natural selection can operate so as to produce intelligent life. This process is exceedingly improbable, given the huge number of other possibilities that would not produce intelligent life.

Dawkins would probably say that any other outcome would be equally improbable, though obviously we would not be around to assess its probability, and probably for most outcomes there would be no life around at all. But what makes the improbability of our planet so interesting is that it has produced such very complex, organised entities. Through a process of basic simplicity (repeated mutation and selection) that has consequences of amazing diversity (the millions of species that have existed on earth) evolution has given rise to conscious, intelligent, morally responsible organisms. If there is great intrinsic value in the existence of such organisms (and maybe a lot turns on that judgement of value), this is just the sort of process that a Newtonian God (a God of great intelligence and cunning) might well devise. So it is hard to see why Dawkins should think that Darwinian evolution should even be in tension with belief in God, much less incompatible with such belief. Darwinian evolution could exist without a plan or purpose, though it would be incredibly improbable if it did. But evolution could well be the elegant plan of a supreme intelligence (just like Dawkins setting up his 'Biomorph' program on his home computer; cf. *The Blind*

Watchmaker, chapter 3). And if it were, it would be much more probable – perhaps even certain – that the process would turn out as it has done, for God would set it up to ensure that outcome. Though the process is possible without a designing intelligence, it is much more probable if there is such an intelligence. I think it is a good principle to take the more probable alternative, other things being equal (which, of course, they are not for Dawkins – but that is another story).

7. ALTERNATIVE FUTURES

Another reason a biologist might have for thinking of the evolutionary process as accidental or blind is that the process could easily have happened otherwise. There seems to be no necessity about how evolution has developed. So it is just chance that it has developed in the way it has. As the American palaeontologist Stephen J. Gould, the best-known exponent of this view, put it, if we ran the story of evolution on earth a second time, things would probably happen quite differently. We might have an earth dominated by some of the long-vanished fossil forms found in the Burgess Shale, instead of by mammals. Or we might have no surviving life at all. Meteorites or comets might have extinguished all life long ago. It is just chance, accident, that humans exist at all. Humans are not the culmination of a developing 'tree' of evolution; they are accidental side-branches, by-products of a totally fortuitous (for us) process.

It is interesting that Gould places such an emphasis on alternative futures. For physical determinists, of course, if we ran the program again, everything would happen in exactly the same way, even down to the impact of various meteorites on the earth. Gould is assuming that, from the same initial condition of the universe, a number of different tracks into the future are possible. If that is so, he implies, nothing but chance could determine which track is taken.

To many biologists this seems like exaggeration. Simon Conway Morris, who worked on the Burgess Shale, points out that 'To mainstream evolutionary biology the existence of trends is entirely unremarkable' (*Life's Solution*, Cambridge University Press, Cambridge, 2003, p. 304), and 'trends imply directionality'. The matter is discussed in detail in K.J. McNamara's, *Evolutionary Trends* (Belhaven, London, 1990). It is clear that many biologists find no difficulty in speaking of trends, or weighted trajectories of development, in the evolution of life on earth. Conway Morris concludes, 'When within the animals we see the emergence of larger and more complicated brains, sophisticated vocalisations, echolocation, electrical perception, advanced social systems including eusociality, viviparity, warm-bloodedness, and agriculture – all of which are convergent – then to me that sounds like progress' (*Life's Solution*, p. 307).

This sounds like an echo of Julian Huxley's remark that nature is 'unitary; continuous; irreversible; self-transforming and generating variety in novelty during its transformation' (Julian Huxley, *Evolution in Action*, Chatto & Windus, London, 1953). Huxley, avowed humanist and grandson of Darwin's defender T.H. Huxley, was quite clear that there is a direction in evolution, whether it is blind or not.

Biology has moved on a long way since Julian Huxley, but it is still plausible to say that 'there has over biological evolution as a whole been an overall trend towards and an increase in complexity, so that it is right to speak of a propensity for this to occur' (Arthur Peacocke, *Theology for a Scientific Age*, Basil Blackwell, Oxford, 1990, p. 67).

This conclusion is supported by the work of physicists like Paul Davies, who speaks of 'the discovery of self-organising and self-complexifying processes in nature' that suggest 'there exists a creative arrow, pointing in the direction of increasing richness, diversity, and potential' ('Emergent Complexity, Teleology, and the Arrow of Time', in *Debating Design*, ed. William A. Dembski and Michael Ruse, Cambridge University Press, Cambridge, 2004, p. 208).

It cannot be pretended that there is consensus on this topic. But at least it is scientifically respectable to see the open tracks into the future that Gould's account supposes to exist at various points in the story of evolution as weighted in a specific direction, that is, towards a goal of value. If that is so, it is misleading to speak of evolution having occurred by pure chance. It can look more like, as Arthur Peacocke has expressed it, an unfolding of 'the divinely endowed potentialities of the universe through a process in which its creative possibilities and propensities become actualised' (*Paths from Science Towards God*, Oneworld, Oxford, 2001, p. 77). We can see in the evolutionary process a combination of a propensity towards complex value and a space (at least in its later stages) for free creative choice by conscious organisms. That in turn looks remarkably like a process designed by an intelligent and creative mind to enable other intelligent and creative minds to exist.

Any form of divine determination of or influence over one possible track as opposed to its alternatives would be undetectable by physical science, since it would be a non-physical causal input. Such an input, which could take the form of a general cosmic vector rather than a set of specific discrete acts, might influence a possible future, weighting an outcome that might, or might not, have occurred by pure chance, since it would lie within the range of physical possibility in any case. The only way we could decide whether or not God was so influencing the process would be to ask if the direction of the process was such that a wise creator could choose it. This is not a scientific question, but a matter of how we evaluate consequences – and here differing assessments can be made of the evidence we have from the course of evolution. One important root of belief in God is the sense that personal life is of intrinsic value and importance. If we have such a sense, it will be reasonable and consistent with modern scientific understanding to see the evolutionary process as directed towards the emergence of personal values, and so as expressing the purposive activity of a creator God.

5

FREEDOM, SUFFERING AND OMNIPOTENCE

Acceptance of an evolutionary worldview forces us to rethink some traditional pictures of God. No longer can God be thought of as a person who creates a perfect universe, who introduces suffering and death as a punishment for human sin and who could eliminate all suffering if God chose. Destruction, competition and suffering are essential parts of the process, from long before the advent of human life, and even an omnipotent God cannot remove them completely. God is still the ground and goal of the process, but has created a universe with its own internal constraints and necessities. Without them, we, as parts of the aeon-long process of evolution, could not exist.

AIM: To show that, though evolution is sometimes seen as wasteful, accidental and horrific, it can also be seen as an elegant structure for producing beings of great and otherwise unobtainable value. The God of evolution is constrained by the necessities of nature, but God has chosen to create nature for a good purpose. To show how this provides a partial explanation of the existence of suffering and sin in a divinely created universe.

1. CONVERGENCE AND FITNESS LANDSCAPES

Within the world of evolutionary biology there are many, as I have indicated, who feel that a direction can be discerned in evolution, whether it is planned by a creator or not. This direction is not such that every organism continually seeks to move towards a higher goal. That would be a naïve interpretation of directionality.

Nonetheless, genes that provide the recipe for organisms with better survival and procreative skills will be propagated more profusely than genes that provide recipes for weak and infertile organisms. If there are frequent mutations in the gene pool, new recipes will be continually generated. Some of them will give survival advantages over others, and they will tend to become dominant.

It is obvious, however, that not all genes mutate so as to improve the survival chances of the organisms constructed from them. In the genome of any surviving organism, we might expect there to be some genes that give a survival advantage to the organisms for which they provide the building recipes, some genes that are no longer particularly advantageous but have given a survival advantage at some time in the past, and some genes that are neutral or even disadvantageous but not sufficiently so to destroy the reproductive capacity of the organisms they produce.

It is not true that all mutations are beneficial, or that all organisms are continually getting better. Many result in a loss of functionality, and so are potentially harmful. Many bacteria have found an evolutionary niche that promotes their survival and enables them to withstand, or even benefit from, the existence of more complex organisms without becoming any more complex themselves.

So Matt Ridley, in his wonderfully illuminating book *Genome* (HarperCollins, London, 1999) insists that 'There is no such thing as evolutionary progress' and that 'the black-smoker bacterium ... is arguably more highly evolved than a bank clerk' (p. 25). The bacterium is well adapted to its little niche in sulphurous vents on the bed of the Atlantic, whereas the unfortunate bank clerk may be prey to all sorts of anxieties and phobias that make his existence insecure. The bacterium is so well adapted that it has no need to change. The bank clerk is so ill adapted that he spends his whole life giving and receiving change.

Yet even Ridley, with his animus against bank staff, has to admit, just two pages later, that there is 'a sort of accidental progress' in evolution. Some things do get more complex, bigger and brainier. At least the bank clerk knows that he is not perfectly adjusted to his environment, and can possibly do something about it – maybe write books about bacteria. From total unconsciousness to intelligence and understanding is some sort of progress, as any teacher will testify.

The fact is that, though not all organisms become more complex, some do, if genetic change enables them to find a higher place in what Sewall Wright, in the 1930s, called a 'fitness landscape'.

A fitness landscape is not an actual physical landscape. It is a diagrammatic construct, which plots the probabilities of various alternative future developments among biological organisms. In such a landscape, there are 'valleys' where lots of simple organisms continue to flourish and there are 'hills' that provide environments where organisms can evolve in various ways by 'climbing' into new niches – until some organisms reach the peak of one possible line of development. Perhaps bacteria inhabit one such peak, since they adapt to their environment admirably. Humans are at or near another peak, where the ecosystem of the earth gives maximal adaptiveness to intelligent sensory organisms.

For this to be so, some genetic mutations must be progressive, in the sense that they are small enough to preserve the organism, productive of more complex structures that build on already developed ones, and able to enhance the adaptive characteristics of the organisms constructed from them.

An example would be the way in which eyes develop (and have done so in many diverse contexts) from variations in skin sensitivity, over many fairly small and continuous mutations. Salvini-Plawen and Ernst Mayr, in a well-known paper in

Evolutionary Biology in 1977 ('On the Evolution of Photoreceptors and Eyes'), claimed that there are over twenty different and independent lines of differentiation that have led to the development of camera-eyes in organisms. There seems to be a convergence, encouraged by the ecosystem, on specific forms of organic adaptiveness. This may well extend, Simon Conway Morris suggests, to the existence of sentience and intelligence.

Although not all organisms are 'improving', the fitness landscape of the earth may ensure that in the end, given suitable sorts of progressive genetic change, there will be some organisms that climb to fill the ecological space available for intelligent agents.

Simon Conway Morris (in *Life's Solution*) has argued that the nature of the earth's ecosystem is such that there are only a finite number of spaces available for organisms to move into. As the spaces fill up, the constraints of biological development force evolution in a specific direction. When one 'valley' is full, sooner or later organisms will produce mutations that will move some of them over the hill into the next valley, or up towards a mountain peak. If one of the mountain peaks is that of conscious intelligent organisms, then when every other space is filled to overflowing, and organisms continue to mutate in all sorts of ways, it becomes virtually inevitable that some intelligent organisms will develop. There is simply no space left for anything else to happen.

This can be seen as an extension of the fine-tuning argument. Because of the atomic structure of carbon, the number and character of available chemical elements, and the relations of these elements to one another, carbon-based life will have a natural tendency to develop some sensory and intelligent forms – even, Morris suggests, so far as bipedal mammals with the sort of sense organs humans have.

In this sense, humans or something rather like them result from natural tendencies implicit in the fine structure of the cosmos from the first. They are entirely natural, and possibly inevitable, constituents of exactly this cosmos. It looks as if there could well be a predetermined directionality in evolution. It even looks as though we humans are, in Paul Davies' words, 'meant to be here'.

2. STOCHASTIC EVOLUTION

Stephen Gould's point about a lack of direction in evolution might still be made by saying that the course of evolution on earth, even if it is the result of a natural propensity, is not such that it could be chosen by a wise and good creator. True, the evolutionary process has perhaps led in accordance with elegant laws to states of intrinsic value, and the laws of nature are indeed mathematically elegant, but they also seem to result in great cruelty, or at least in great amounts of pain. There are many states of intrinsic value among organic life-forms – states of great happiness, of understanding of the environment, and of appreciation of the beauty of the world. But mutations are as often harmful as they are beneficial, and whole species have been exterminated without leading anywhere. So it seems there is no predetermined line to humans in the evolutionary process, that the process has been extravagantly wasteful of life and has also caused great suffering, as species have preyed upon each other: nature has seemed 'red in tooth and claw', as Tennyson put it.

The situation seems to be that the fundamental physical laws are simple and elegant, and they lead by processes of stable and cumulative self-organisation to complex structures in which intrinsic values are actualised, but those processes also produce outcomes that are destructive, and some such outcomes seem essential to the processes' operation. We can see this necessity of destruction very easily. The thermonuclear destruction of stars is necessary to release carbon, the basis of all organic life, carbon being manufactured from helium inside large stars. The death of many simple species is necessary to the generation of more developed life forms. And life grows only by the consumption of energy and matter, which usually involves the destruction of other life-forms.

Evolution could not occur without mutations of DNA. Such mutations often have no consequences for the fitness of organisms to survive. Often they are harmful to organisms. But on occasion there are enough favourable mutations to drive the process towards greater adaptiveness and integrated complexity.

I suppose we could imagine a God who determined each mutation so that it only improved organic structures. Such a God would be involved in the natural order in a very determinative way. Everything would be exactly what God wished it to be. There would be no room for creative spontaneity or autonomy in the created order.

Perhaps there are universes in which God does unilaterally determine everything for the best. Any conscious beings in such a universe might be perfectly happy in being moved entirely by God to ever new levels of enjoyment. Some people have thought the universe we inhabit is in fact like that. But if it is, God does some very strange things, in directly causing people to commit great evil and undergo great sufferings.

The universe does not look as if it is determined in every respect by God for the best. If there are going to be general laws of physics, and laws that involve some element of destruction in order to engender new forms of life, then God cannot both keep those laws and make sure that nothing unfortunate ever happens. For example, if there is a general law of gravity, God cannot ensure that no one ever gets hurt by falling from a high mountain, unless God breaks that law whenever someone is in danger. A universe in which God was continually suspending the laws of nature to save everyone from harm would be very different from this one. No one would ever get hurt. But there would be no predictable and reliable laws of nature either. So if God opts for a law-like universe, it is impossible for God to determine everything for the best.

In addition, there are many who think that radical freedom is a necessary condition of moral responsibility and of there being persons in this universe who are to a large extent self-moving, autonomous and creative. If persons are autonomous in the sense of being radically free, nothing – not even God – can completely determine how they act. They can at least to some extent choose for themselves how they will live. They can choose one out of a range of tracks into an open future, without being determined to do so by God or by anything else. God gives them the power of choice, and sustains them in existence when they have chosen. But the choice is theirs, and they must live with the consequences for good or ill.

In such a universe, God cannot unilaterally determine everything. At least some tracks must be left undetermined. There must be open, undetermined futures built into the very structure of the universe. If so, mutations in the process of evolution will sometimes be undetermined. And if nothing is determining them for the best, they will be 'random'. That is, they will often be harmful to organisms, just as they will often be advantageous to them.

Naturally, the future cannot be completely open. There will be limits placed on what is possible, and we would not expect God to leave the future entirely to the mercy of human beings or other creatures. We might expect God to keep overall control, but to allow autonomy of choice within the limits needed to keep such control. We might well think that God will ensure that the main divine purposes get realised somehow, but will allow various ways of achieving them. God, in allowing freedom, will also allow choices that obstruct the goals of creation, but only to a degree that can be rectified later in some fashion. We might say that in such a universe many physical processes will be stochastic (that is, statistically, but not completely, determined). They will allow details to be filled in either randomly or by the creative choice of many finite agents, but the general pattern and the ultimate goals of the cosmos will be divinely determined.

3. EVOLUTION AS DIRECTIONAL

As I have noted, evolutionary biologists are not agreed about the extent to which the evolutionary process is determined. All would, I think, agree that the process is normally governed by general physical laws, rather than by direct divine actions that ensure that everything is for the best. This means that God did not create an absolutely perfect universe which went wrong at a later stage because of human sin. But some biologists think the laws themselves (together with a set of initial conditions) might totally determine all events, so that everything that happens is bound to happen. There will be no radical freedom and no possibility of alternative tracks of evolution.

On the other hand, it looks to other biologists as though there is a structure of law that has led to the generation of intelligent life-forms, but has left many details of the process to be filled in either randomly or by a combination of creative decisions made by many creatures. Creative decisions probably only become relevant quite late in the process of evolution, after some degree of intelligence has begun to exist. Before that, it will be more chance or sheer indeterminism that drives the process of evolution.

You might think that the determinist view is the one most sympathetic to belief in God, since then you can see the whole process as predetermined from beginning to end. God's purposes are bound to come about, because God has designed the laws that determine everything that will happen.

The indeterminist view seems at first sight to make the process too random and unpredictable, so that even God is not in complete control of the universe God has created. However, indeterminism seems to be a necessary condition of radical freedom. It makes radical freedom and genuine creativity possible. At first, before there are any conscious intelligent agents, it may just look like chance. But once such agents come to exist, it can be seen that having different tracks into an open future is precisely what is needed to allow finite agents to be really creative and responsible.

In such a universe, intelligent creatures will not be wholly determined by a supreme intelligence. But neither will they be wholly free to do anything they wish in a universe without any structure, direction or goal. They will be free only within limits. God can still have a purpose for the universe, and God can ensure that the general divine intention is realised, sooner or later.

We might imagine a great intelligence setting people on an island and telling them that they must travel to find a desirable place to live. They can choose any direction in which to travel, and do anything they wish on their journey. But this intelligence knows where the jungles, caves, swamps and dangerous animals are on the island, and so can predict very well the sorts of things that will happen to all those people. Also, the intelligence knows that at the far end of the island there is a beautiful city in which each person can find whatever they like best. Even the supreme intelligence cannot tell exactly what journey each person will make, what dangers they will encounter and how they will react to them. But it can be confident that a great many of them will find their way to the beautiful city and remain there – because that, in the end, is the only place they will be safe, happy and fulfilled.

So God could create a universe in which there is much room for creative freedom and for the exercise of moral responsibility, and yet have a goal for the universe, which God could predict would be achieved at least by many. The future of each agent is open, but the deepest nature of the universe sets a goal of perfection that is bound to be realised in the end. (This possibility is explored by David Bartholomew in *God of Chance*, SCM, London, 1984.)

This is a possible universe. Is our universe like that? From a purely scientific point of view, we cannot be sure. But it seems to be consistent with the facts of evolution. What can from one point of view seem to be a profligate wastefulness in nature can also be seen as an exploration of various tracks into the future in a variety of environments, some of which must fail but some of which will succeed. What seems to be random mutation in fact generates a highly integrated flexibility and adaptability of organic forms that can discover adaptive fitness with great efficiency (cf. Peacocke, *Paths from Science towards God*, chapter 4). So the combination of a number of elegant physical laws together with a limited degree of indeterminism or randomness makes possible a universe that necessarily develops, in partly unpredictable ways, towards the goal of intelligent consciousness. When creaturely freedom enters the picture, it takes over what were random occurrences and transforms them into events of creative choice. The whole structure is overseen by the intentions of God, which do not interfere with the created structure but guide it in an empirically imperceptible but causally powerful way to its goal.

4. THE PROBLEM OF WASTEFULNESS

Seen thus, the process is not wasteful. It is a finely balanced harmony of forces of destruction and creation, of regularity and novelty, of conservation and change, which must operate together if the universe is to develop its own inherent potentialities and so transform itself from material simplicity to intelligent complexity.

It would not be true to say that God was absent from this process, since it only exists by the sustaining presence of God. But God could exercise a hidden, divine causality, unobservable and untestable, in order to allow the possibility of cosmic self-unfolding that is the history of this universe.

The universe grows to unfold its own inherent possibilities, as a lotus flower unfolds from the seed. God is like the sun that draws the plant to the light, but allows it to be all that it can be. Intelligent beings can encourage and direct this growth, or impede it, though only for a while. There is no real waste in this process, although to make sense of what happens we have to see it in the context of the growth of the whole. The process will come to an end, and what has grown will die and be no more. Yet it will have been, and its reality will be remembered in the mind of God for ever, as a thing of unique beauty that would not otherwise have existed.

From God's point of view, the dinosaurs were not a waste, even though they became extinct. Everything will die, and it is estimated that the dinosaurs flourished for at least 150 million years before their death. The value was not that they continued for ever, but that they once existed, and that existence is not a waste just because it did not lead directly to humans.

5. THE PROBLEM OF SUFFERING

Suffering, like pleasure, is part of the process of increasingly sensitive organic response to stimuli from the environment. Until the formation of fairly large brains, suffering probably does not take the agonizing form it does in humans. So the thought of ichneumon flies eating live caterpillars from the inside does not possess the horrifying aspect that it may at first seem to have – and that it seems to have for Stephen Gould. Biologist David Hull has described the evolutionary process as 'rife with happenstance, contingency, incredible waste, death, pain and horror'. A great deal of this 'horror' may be due to an anthropomorphising of the natural world, reading into the inner lives of diverse animals the sorts of suffering humans endure. If the occurrence of suffering as humans experience it is closely connected with the release of chemicals in a highly structured brain, then it seems likely that insects and even mammals with less developed brains may not experience suffering as we do. They react by aversion to unpleasant stimuli, but perhaps without consciousness, or with forms of consciousness that are dream-like and less emotionally charged than those of the higher mammals. If this were so, it would remove most of the horror from the evolutionary process.

Yet the higher mammals, at least, suffer pain, and that may seem hard to square with the idea of a benevolent creator. This, of course, is far from being a problem produced by evolutionary theory. The problem of pain has always been a problem for theists. But a study of biology suggests that pleasure and pain are produced by the sensitivity of the nervous system to stimuli, and they are interconnected in complex ways.

Pain is a possibility rooted in the creator's mind that is the basis of all possibility. It is not wilfully chosen. It is necessarily there, as possibility. We might think of the spectrum of pain and pleasure as something like the colour spectrum, with a range of possible colours that are actualised when wavelengths of specific amplitude occur. So there is a pain–pleasure spectrum, and parts of it are actualised when stimuli of specific sorts impact upon the developed brains of organisms.

6. THE PRIMORDIAL SACRIFICE

What causes the problem is that we think of God as a person who chooses to cause pain, when God easily need not have done so. If we think instead of central nervous systems developing various modes of response to the environment, we can see how aversion leads to subjective feelings of pain, and attraction leads to subjective feelings of pleasure, by natural and perhaps inevitable processes of development. Pleasure and pain as we experience them are necessarily involved in the possibility of this universe and will be actualised under specifiable and sometimes unavoidable conditions.

This means we have not to think of God as able to stop absolutely anything at any time. Pleasure and pain are potentialities that become actualised in the course of the self-development of this universe from the ground of their possibility in God, and there is nothing God, or anyone or anything else, can do about them.

But, it may be objected, is God not omnipotent? Can God not do absolutely anything God wants? I think that is far too anthropomorphic a view of God. We imagine a being that can do absolutely anything – like creating a universe of conscious physical beings evolving by natural selection without any pain at all – and presume that such a being could really exist. But how could we know this? We have no idea of what a supremely intelligent mind would be like and what constraints there might be on what it could do.

We can say that God is omnipotent, if we mean that there is no possible power greater than that of God, and all power derives from the being of God. Such a being would be the most powerful being there could ever be, and there could be no power that could oppose it or destroy it. What more could we want? Yet such omnipotence might not be able to change absolutely anything. It could not, for example, change its own essential nature, and in that nature are rooted all the interconnected possibilities of being that are necessarily what they are.

Perhaps this is the thought enshrined in so many ancient religious myths about the universe coming into existence through the primordial sacrifice or self-immolation of a God. Pain is involved in the emergent physical being of universes like this, and even God, who experiences all actualities, must share it if such a universe is actualised. If we do not start from some imagined idea of an all-powerful person, but from the idea of an unknown reservoir of potentiality that expresses its being in this physical universe, we might have a more adequate starting point for thinking about God, the ultimate mind of the universe.

7. ASSESSING EVOLUTION

It looks as if the universe is directed to the actualisation of great values (like understanding, beauty and love), but only through a process that necessarily entails the existence of much suffering. The beauty, the elegance, the desirability of the ultimate goal (the goal of values actualised in the world and remembered for ever in God), all suggest a supreme intelligence concerned for the realisation of value. But suffering and destruction suggest the existence of necessities, rooted in the very nature of primordial being, which cannot be evaded. The God of the scientific worldview seems like a supreme intelligence aiming at the actualisation of ultimate values through the generation of autonomous structures that are constrained by their own inner necessities.

Ancient religious traditions have sometimes accounted for suffering by saying that it results from the evil choices of finite wills – perhaps Satan, perhaps a primal human being, Adam, or perhaps souls who have fallen from the stars by desire for selfish pleasures. An evolutionary worldview rather suggests that suffering is inherent in the structure of pre-intelligent nature, and is part of the situation in which intelligent beings find themselves existing from the first. That means that suffering is not a punishment for sin, but is a necessity of the natural order, which has to be faced and overcome or in some way transcended.

The truth in the ancient myths is that human wrongdoing undoubtedly makes the degree and extent of suffering much worse than it otherwise would have been. But in the absence of wrongdoing, earthquakes would still have occurred, volcanoes would have exploded, viruses would have replicated and destructive accidents would have happened. So evolutionary science does throw doubt on a certain traditional conception of God – that God chooses exactly everything that exists and could eliminate pain from this universe if God chose. It suggests instead that if God 'chooses' this universe, God has to accept the pain that goes with it, and there is nothing that even God can do about it.

We may say that God need not have chosen this universe, but then we would not have existed at all. If God chooses us to exist, then God must choose this universe, and must choose it as a self-developing holistic cosmos, in which there are strong interconnections between the parts – interconnections between pleasure and pain, for example – that even God cannot change.

8. THEISTIC EVOLUTION

For some people, evolutionary thought suggests that the complex can emerge from the simple by a blind and automatic process. The many blind alleys and the amount of suffering in evolution count against the existence of any intelligent creator, and the principle of economy means that we should not introduce a God if we do not need to do so. From this perspective, Darwinism undermines the argument that there is design in the universe.

However, the process does not have to look blind and automatic. Charles Darwin himself never gave up some sort of belief in God. While writing his autobiography, in 1879, Darwin wrote, 'In my most extreme fluctuations I have never been an atheist' (quoted in Alister McGrath, *Dawkins' God*, Blackwell, Oxford, 2005, p. 76), although he confessed that he sometimes felt like an agnostic. But to him the thought that the laws governing evolution were somehow designed seemed overwhelming. And to many biologists, as we have seen, the evolutionary process at least *looks* as if it is directed towards the existence of intelligent consciousness. The process seems to move by a series of improbable but mathematically elegant steps towards intelligent consciousness as its inevitable outcome. Moreover, it exhibits a flexibility and spontaneity that suggest a riotous creativity and energetic exploration of possible tracks into the future, an inner vitality that speaks of continual striving to advance into all possible niches for life. It is more like a dynamic striving to new levels of complexity and organisation than a grinding out of repetitive patterns characterised by constant copying errors that result in intelligent life by pure accident.

Maybe Erasmus Darwin got it right. At least his leading metaphors were not the pessimistic and negative ones that have become associated with neo-Darwinism in our own day. Instead of a ruthless competition of all against all, we could see evolution as a co-operative building up of ever more complex unities, and the conflicts that occur in the process as necessary stages in the development of new forms of life. Evolution could then be seen as the vital struggle of insensate nature to develop new, more sophisticated, intelligent and adaptive forms of being out of the death and transfiguration of the old. It would be the unfolding of potencies grounded in a vast intelligence, but with their own principles of development and their own unique character, the actualisation of possible worlds that are emergent and creative wholes.

The evolutionary perspective that is such an important part of our modern scientific worldview does appear to rule out the idea of a benevolent God who can do absolutely anything and who is always making everything turn out for the best. For some, that is enough to rule out the idea of God altogether. Then a Darwinian account provides a good enough explanation of how the complex can emerge from the simple by many repeated steps, and how an appearance of design can result from a blind process of mutation and selection.

Yet the whole process is still amazingly improbable, and it is puzzling that apparently free, intelligent, conscious life results from such a completely unplanned set of genetic copying errors. So for many evolutionary biologists, including Charles Darwin himself, the most probable hypothesis is that there is some direct-ive intelligence that has set the system up precisely in order that conscious, self-directing life should evolve. This intelligence is not benevolent in the sense of removing all suffering from the universe. And it is not very clear what its purpose might be – especially if you cannot see much point or value in human life. But if you think personal lives are of immense value, and that conscious life is something to be prized, you might well think that evolution, far from contributing to the death of God, supports the idea of a supreme intelligence bringing into being a universe that can, as Charles Darwin said, 'tend to progress towards perfection'.

9. DOES GOD EVOLVE?

The image of a universe progressing towards perfection is so plausible, given a general evolutionary picture of the cosmos, that it has led some – like the now almost forgotten early twentieth-century British philosopher Samuel Alexander – to suppose that the universe is itself gradually evolving into God, into a vast cosmic supermind that will in time be omnipotent and omniscient, but has to struggle though aeons of sometimes painful and always difficult development to become so.

There is, however, something odd in the thought that an omnipotent and omniscient being should come into existence by accident. There is also something odd in the thought that these 'accidents' happen in a universe whose basic structure, from the very first, exhibits incredible mathematical elegance, beauty and intelligible rationality.

It seems incredible that a cosmos that begins blindly, unconsciously and randomly should slowly become divine. It is as if God says, 'Luckily for me, I am omnipotent and omniscient. But that, of course, is just an accident.' The Catholic palaeontologist Pierre Teilhard de Chardin, who talks of the universe moving towards an 'Omega Point' of self-conscious intelligence, has been accused of propounding such a view. But he makes it clear that he considers 'Point Omega' to have always existed and that the cosmos moves towards unity with that pre-existing reality (Teilhard de Chardin, *The Phenomenon of Man*, ed. Bernard Wall, Collins, London, 1959). This is indeed a more plausible conjecture. There is from the first, or indeed timelessly, a supreme cosmic intelligence that contains within itself the potentiality for every possible universe that could ever be. It could generate from these possibilities an independent physical cosmos. That cosmos could evolve from a first state of almost total potentiality – the 'big bang' – a state of unconscious dynamic potentiality. That potentiality contains negative, destructive elements as well as positive, constructive ones. Some negative elements are bound to be realised as the cosmos moves towards increasing self-awareness and self-organising realisation of its inherent nature. Under the guiding but not determining influence of the cosmic mind, such a universe would actualise new forms of value and new forms of conscious agency that would be unique to this universe. Through a combination of chance and necessity, guided by infinite intelligence, the evolving cosmos shapes its own future, and thereby actualises what was only potential in the divine being but now comes to have an actuality that exhibits both spontaneity and intelligibility in a unique way.

That is a picture that evolutionary theory can suggest. It revises some traditional ideas of God, but it is not so far away from some of the ideas of Plato in his Dialogue *Timaeus*, which had a profound influence on the development of medieval Christian ideas about God. And it sees evolution, not as a blind process arriving by accident at conscious personal life, but as an expression of the creativity of the mind of the cosmos, destined from the first to generate societies of finite intelligent beings that can have some responsibility for the future of themselves and of their world.

6

THE VEILED WORLD

Quantum physics has destroyed the old materialist picture of the physical universe as composed of solid particles banging into one another. For many quantum physicists, the physical world is a 'veiled reality' that we can only know as appearance, as it appears to us, not as it is in itself. For some, the quantum world of wave functions in Hilbert space points to consciousness, or to a 'Platonic' world of conceptual forms, as a primary element of reality. These interpretations of quantum physics enable us to think of the real basis of the material world as an intelligible realm that exists in the mind of the primary consciousness, God. That makes God not an optional add-on to the universe, but essential to the very existence of matter itself.

AIM: To show that quantum physics undermines old-style atomistic materialism. To show that the mind of God might enter into the very structure of physical nature in an intimate and essential way. To show that the observed world is an appearance of a deeper, veiled reality.

1. THE QUANTUM REVOLUTION

It is possible to interpret the Newtonian and Darwinian revolutions in a materialistic way, although that would not reflect the views of Newton and, admittedly to a lesser extent, Darwin. But the fourth great revolution brought about by modern science seems to undermine materialism in a dramatic and unexpected way. It certainly undermines the sort of materialism that sees the universe as consisting of solid material particles that interact with each other in a mechanical, bump-and-shove way.

The fourth revolution is so different from what went before it that it is sometimes called 'the new physics'. It is the physics of quantum theory. There is no one 'great name' that can be associated with this revolution. Einstein at first comes to mind, and in 1905 he did write a ground-breaking scientific paper – one of four papers he published that year – that helped to lay the foundation for quantum physics. However, Einstein is more properly seen as the founder of the theory of relativity – another of his 1905 papers laid the foundation for that. Quantum physics was developed by a number of other physicists, without one name being of obviously supreme importance.

Einstein's contribution was his paper on the photo-electric effect, a study of experiments in which a beam of light knocks electrons out of a plate of metal. He proposed that light consists of small particles (photons), and not just of waves, as was generally thought. (There are many good introductions to quantum theory. A readable one is Paul Davies and J.R. Brown's *The Ghost in the Atom*, Cambridge University Press, Cambridge, 1986. Another is Nick Herbert's *Quantum Reality*, Rider, London, 1985.)

The particle nature of light was confirmed by Robert Millikan, who spent years trying in vain to disprove it, and it is very important for the development of quantum physics. Werner Heisenberg devised the first quantum theory in 1925, and Erwin Schrodinger and Paul Dirac developed theories that turned out to be equivalent, though using different mathematical symbolisms.

Quantum physics in fact deals with a physical reality that seems to be both particle-like and wave-like at the same time, however impossible this may appear. The photo-electric effect shows light to behave in a particle-like manner, and the two-slit experiment, set up in its original form by Thomas Young in the early nineteenth century, shows it to behave as if it consisted of waves. Light seems to travel as waves and to arrive as particles. To cope with this peculiar situation, quantum physics has to abandon the old materialist perspective of a universe of solid and separate atoms endlessly knocking into one another. Instead of electrons as little solid planets circling in defined orbits around the central nucleus of the atom, electrons are seen as consisting of waves (but very peculiar waves, waves of probability) that collapse into particles of definite position and velocity only when 'observed' or measured.

Richard Feynman said that 'no one understands quantum mechanics'. That is very comforting for people like me. But he did not mean that the mathematics or the physics cannot be understood. There is no doubt that quantum mechanics is immensely successful at prediction. In practice it has been comprehensively confirmed, and is the basis for most of our computing and information technologies. It works. But what the theory might correspond to in objective reality is baffling, and there is no agreement among physicists on how to understand what it tells us about the basic structure of physical reality. There is little doubt, however, that it undermines the theory that the physical world is made up of small particles that interact locally (by contact) in accordance with deterministic and completely knowable laws.

For quantum theory, there seems to be a complete difference between what the physical world is like when it is not being measured and what it is like when observed. When not being observed, it is best described as a set of probability waves. To be specific, the square of their amplitude (the height of the wave from top to bottom) gives the probability that, if measurements were made in a specific way, a particle would be found at a particular location. But the observer has a choice about what is to be measured, and waves can be described mathematically in a number of ways, depending on the sorts of measurements or observations that are of interest.

A measurement involves setting up an experimental situation in which the impact of particles is recorded on a phosphor screen or other recording device. The experimental apparatus constrains waves to 'collapse' into particles, the location of which can be predicted with a precise degree of probability. Waves themselves are never observed. We humans only see particles, or the evidence that particles have impacted on a screen or recording device in a specific way. We might therefore say that the apparatus selects for us, out of wave-forms that we cannot perceive, particle impact patterns that we are able to perceive. At a specific point in time, the apparatus can freeze these patterns on some recording device, like a photographic plate, so that we can observe them later. The patterns that are chosen are precisely ones that we humans can perceive. So we might say that particles like photons or electrons are posited as frozen selections from wave-forms, selected because they are suitable for human perception.

SUGGESTIONS FROM QUANTUM THEORY:
A. THE PRIMACY OF CONSCIOUSNESS

2. WAVES AND PARTICLES

Which are real: waves or particles? On this opinions are divided. Most working quantum physicists do not get too concerned about the metaphysical implications of quantum theory. Among those who do, views range from Niels Bohr's 'Copenhagen interpretation', that the only reality we can describe consists of observed phenomena and that talk about probability waves is a purely mathematical construct, to John von Neumann's claim that the fundamental physical reality is one of probability waves in infinitely dimensioned Hilbert space, while the phenomena we see exist only in the mind. In between lie less popular views like that of David Bohm – the physical world consists of particles after all, but with very peculiar and decidedly non-Newtonian properties – and Hugh Everett's extraordinary claim that every possible quantum world actually exists (a readable popular account of these interpretations can be found in Nick Herbert, *Elemental Mind*, Plume New York, 1994, chapter 5).

Underlying these very different interpretations is the fact that what humans actually perceive in laboratory experiments are particles, or the impacts of particles. Waves are postulated to account for the patterns such impacts make. So, though some theorists affirm that probability waves really exist, other physicists have a preference for particles, which at least are actualities, not just probabilities. These latter workers then view probability waves as purely theoretical devices useful for calculations but not corresponding to 'real' waves in an objective world at all.

But either preference carries with it some unusual implications, very different from those of classical physics. For it seems that particles only really exist when they are measured. Niels Bohr and Werner Heisenberg, who in his later writings came to agree with Bohr, take this to mean that we can only speak of interactions between observers and objective physical reality. Everything else – what reality is 'in itself' – is unknown. So the physicist does not, as in a simple Newtonian picture, describe physical reality as it truly, objectively, is. The quantum physicist chooses to interact with the physical world in a specific way and records only the results of such interactions. The Newtonian world of solid bouncing particles dissolves into a much fuzzier, almost wholly opaque, reality, which we can only know insofar as we interact with it. The ways in which we interact with it largely determine what we perceive it to be.

Some – a minority, but a significant one – go much further than this. John Wheeler holds that no elementary phenomenon is a real phenomenon until it is an observed phenomenon – 'It has not really happened, it is not a phenomenon, until it is an observed phenomenon' ('The Past and the Delayed-Choice Double-Slit Experiment', in *Mathematical Foundations of Quantum Theory*, ed. A.R. Marlow, Academic Press, New York, 1978, p. 14). Philosophers will be irresistibly reminded of the eighteenth-century Anglican Bishop Berkeley's dictum that 'to be is to be perceived'. Nothing is real, the Bishop held, unless it exists in the mind of some observer, whether it is some finite spirit or the mind of God.

Known as Idealism, this philosophical view has been unpopular in recent times, partly because science seemed to suggest that nothing exists except material particles and that the mind is no more than an accidental by-product of the material brain. In a totally surprising way, quantum physics is taken by some to show that Berkeley was more or less right, after all. Nobel Laureate Eugene Wigner holds that human consciousness is directly involved in the collapse of the wave function, the quantum measurement. He writes: 'The very study of the external world led to the conclusion that the content of the consciousness is an ultimate reality' (see Eugene Wigner, 'Remarks on the Mind–Body Question', in *Quantum Theory and Measurement*, ed. J.A. Wheeler and W.H. Zurek, Princeton University Press, Princeton, 1983, pp. 168–81). Particles only exist when observed, he suggests, and so the reality of particles entails that consciousness is a fundamental element of reality, not just a by-product of some 'real' material world.

Raymond Chiao, Professor of Physics at the University of California, Berkeley, points out that Wigner's emphasis on *human* consciousness leaves us in difficulty about quantum events that occurred before humans existed. But quantum theorists are not deterred by such difficulties. John Wheeler proposed that humans, by being able to look back and observe the beginning of the universe, are somehow involved as participants in the origin of the universe ('Law without Law', in *Quantum Theory and Measurement*, pp. 182–213). It seems a much less fantastic hypothesis, says Chiao, that it is the consciousness of God that actualises the physical reality of the universe (Chiao, 'Quantum Nonlocalities', in *Quantum Mechanics*, ed. Robert Russell, Philip Clayton, Kirk Wegter-McNelly and John Polkinghorne, Vatican Observatory Foundation, Vatican City, 2001).

John von Neumann goes so far as to suggest that no observed phenomenon would be real unless consciousness exists – which is a complete reversal of materialism. But he does think that unobserved entities – namely, probability waves – are real. And most quantum physicists suppose that some non-conscious reality underlies the phenomena we see, whatever that reality is like. The true reality is forever hidden, a 'veiled reality', as Bernard d'Espagnat, former director of the Laboratoire de Physique Theorique et Particules Elementaires at the University of Paris, puts it. 'Consciousness as I think of it', he writes, 'cannot in any way, for straightforward reasons of logical coherence, be reduced to a mere property of matter: for it is consciousness which in a sense carves out the atoms within the body of reality' (Bernard d'Espagnat, *Reality and the Physicist*, Cambridge University Press, Cambridge, 1990, p. 214).

This is a more moderate form of Idealism, sometimes associated with the philosopher Immanuel Kant. Niels Bohr admitted the influence of Kant's philosophy on his interpretation of quantum theory. What we know and experience of the material world, for this interpretation, is largely a product of our consciousness. But there is a 'hidden' reality underlying these appearances.

A number of quantum physicists agree with this view, even though we should not expect all physicists to agree about anything quite so theoretical in this area. After all, they could take Feynman's advice and simply not get into this sort of problem. Most quantum physicists probably do not regard consciousness as particularly important in the collapse of the wave function. It may be a purely physical phenomenon, produced by a specific measuring apparatus, which consciousness somehow observes. Yet the measuring apparatus is chosen precisely because it gives rise to events we humans can observe. It makes a mysterious physical reality observable and is in that way closely related to consciousness.

So for some of those who are prepared to wade in deep metaphysical waters – and they include some leading quantum theorists (von Neumann, Eugene Wigner, Henry Stapp, Freeman Dyson) – consciousness becomes of fundamental importance, as what gives form to reality as it appears to us.

For such a moderate form of Idealism, the world in itself, almost completely unknown to us in its essential character, must be distinguished from the world as it appears, as phenomenon. Consciousness, and its specific form, makes a difference to how things are perceived to be, and so consciousness is a constitutive element of reality, not just some by-product of a solid material world of real elementary particles.

An analogy is with perceived colour. Objects have no colour when they are not being observed, for colour arises when wavelengths of light reflected from objects impinge on the eye and coded information is transmitted to the brain. Objects have properties that give rise to sensations of colour when observed, but colour is not an intrinsic property of objects. So in the unmeasured quantum world there are no particles with precise dynamic attributes, such as position or momentum. But on this interpretation of quantum theory, probability waves, whatever exactly they are, generate such particles when they are observed in a specific way, or when they are fixed in time by an experimental apparatus that will give a precise position or momentum when observed.

3. THE WORLD AS APPEARANCE

John Wheeler suggests that the observer actually creates phenomenal reality. A less extreme view would be to say that the observer, and the nature of the observer's consciousness, constructs the reality we perceive out of an underlying reality whose true nature must be forever hidden from us. Either way, it looks as if consciousness may have a fundamental and ineliminable place in our conception of what the physical world is like. In other words, the physical is simply not there, apart from consciousness. Consciousness has to exist for physical reality to exist in the way it does, in relation to us – and we cannot get beyond that to a deeper reality, except in a purely mathematical sense. It is not just 'secondary qualities' like colour and smell that only exist in relation to a human observer. Now the very electrons and atoms out of which physical reality appears to be constituted only exist in relation to a human observer.

If we do get a hint of a deeper and more objective reality, it is by way of the pure mathematics of wave functions and Hilbert space that cannot be mapped on to any world that we can picture imaginatively. Perhaps the underlying reality is purely mathematical, or more conceptual than material in any imaginable sense. Perhaps the very idea of brute matter, and therefore of materialism, must be given up, and replaced by the idea of a deeply mathematical or intelligible world, which appears to us as a physical and phenomenal world – though that is only an appearance, partly constituted by our forms of perception.

On such an interpretation, there is an objective reality, but it is forever hidden and it is deeply mathematical in character. It is remarkably like what Plato called the world of 'Forms' or 'Ideas'. The Forms are non-material, and they function as something like patterns or ideals to which things in the physical world approximate. Whatever this reality is, it is not material, in any ordinary sense. Roger Penrose, whose book *Shadows of the Mind* contains an original and illuminating treatment of the mathematics of quantum theory, asserts the existence of such a 'Platonic world of mathematical forms' as a world that is at least as real as the physical world (*Shadows of the Mind*, Oxford University Press, Oxford, 1994, cf. p. 412 following).

Physical things, for these theorists of quantum physics, dissolve into appearances. Penrose says, 'To me the world of perfect forms is primary (as was Plato's own belief) – its existence being almost a logical necessity – and *both* the other two worlds [the world of conscious perceptions and the world of physical reality] are its shadows' (ibid., p. 417).

Such 'perfect' intelligible Forms, perhaps the basis of the 'hidden' world of quantum physics, might themselves be realities that exist in some form of consciousness. The reason for thinking this is that the intelligible world is a fundamentally mathematical or conceptual world. If we hold, with most mathematicians, that mathematics is in some sense a construct of minds, and if mathematical truths are objective, if they exist apart from any human mind, then the natural conclusion is that they are constructs of a non-human, objective mind, the mind of God.

Such a hypothesis will appeal to anyone who thinks that, on the one hand, the truths of mathematics are constructed by mental activity and that, on the other hand, we discover mathematical truths and do not simply invent them. Disputes about whether mathematics is objective, whether mathematical truths are invented or discovered, and whether the real world is more mathematical than material, cannot be decisively resolved just by appeal to quantum physics. That physics needs to be interpreted, and there are varying ways of interpreting it. Yet all quantum physicists would agree that the Newtonian worldview of massy particles in local interaction has been superseded. In this sense, it is certain that old-style atomist materialism is dead. This does not by any means lead directly to belief in God. There can be other much more sophisticated scientific accounts of the universe that do not refer to an underlying mind or consciousness (the 'many worlds' hypothesis is just one of them). Nevertheless, quantum physics opens up the possibility of understanding mind and consciousness as much more integrally involved in the basic structure of physical reality than anyone might previously have suspected.

Some major quantum physicists think the best interpretation is one that sees ultimate reality as non-material, as an intelligible or mathematical realm. I have suggested that this can be seen as pointing to an objective mind or consciousness that 'constructs' such mathematical truths. As Heisenberg put it, 'Atoms and elementary particles ... form a world of potentialities or possibilities, rather than one of things or facts ... atoms are not things' (cited in Herbert, *Quantum Reality*, p. 195). Again, 'The atom of modern physics can only be symbolised by a partial differential equation in an abstract multidimensional space. Only the experiment of an observer forces the atom to indicate a position, a colour and a quantity of heat. All the qualities of the atom of modern physics are derived, it has no *immediate and direct* physical properties at all' (*Philosophic Problems of Nuclear Science*, trans. F.C. Hayes, Pantheon, New York, 1952, p. 38).

If Heisenberg is right, atoms are slices of probability waves, selected because they are observer compatible. And probability waves, like mathematical entities, are more like mentally envisaged realities – more like possibilities or tendencies – than like spatially localised and solid independently existing elements. In that sense, quantum theory is at least compatible with and makes possible a reconstructed Berkeleyan view that all things ultimately exist in the mind of God. On that view, God is not just the designer of the universe; God is its very foundation and ultimate reality.

7

THE OPEN FUTURE

The dominant interpretations of quantum theory are probabilistic – they affirm that the same cause can have different effects. If this is also true at macrocosmic levels, it allows for radical freedom. Physical determinism is not an axiom of science, and quantum theory shows that it cannot be scientifically established. A picture of the cosmos as creatively moving towards an open future yet fundamentally expressing a hidden but intelligible reality is one that is consonant both with quantum physics and with a basic religious vision of reality.

AIM: To show that creative human freedom and responsibility are consistent with a scientific worldview and with the existence of laws of nature.

SUGGESTIONS FROM QUANTUM THEORY:
B. INDETERMINACY

1. A PROBABILISTIC UNIVERSE

Not only is the real world, the unobserved world of wave functions that seems to be revealed by quantum physics, not made up of discrete and atomistic (separable) parts; it appears to show forms of causality that are non-deterministic. The most widely accepted interpretation of quantum mechanics – the Copenhagen interpretation, named after Niels Bohr of Copenhagen – is that there are genuine indeterminacies in nature. It is not just that we cannot predict the future. Some events happen, at the sub-atomic level, which are not completely determined by any previous physical state or by any physical law.

Abner Shimony, former Professor of Physics and Philosophy at Boston University, writes, 'If the quantum state of a system is a complete description of the system, then a quantity that has an indefinite value in that quantum state is objectively indefinite, its value is not merely unknown by the scientist who seeks to describe the system. Furthermore, since the outcome of a measurement of an objectively indeterminate state is the complete bearer of information about the system, the outcome is strictly a matter of objective chance – not just a matter of chance in the sense of unpredictability to the scientist' ('The Reality of the Quantum World', in *Quantum Mechanics*, 1988, p. 5).

Objective chance, on this interpretation, is not just a limitation on human knowledge. It is a feature of the objective world. So a wave function, which gives a complete description of a quantum system, cannot predict the position of an electron exactly, but only with a mathematically assignable probability. An electron will probably be observed at a specific location, but there is a finite probability that it will not. The reason we cannot predict exactly where it will be is that there is nothing to determine precisely where it will be. The future is open – though within very precisely definable limits – and the electron may take one of several possible paths at random (or in quantum formalism it may take all of them).

The Schrodinger equations, which predict the probable outcomes of particle experiments, are deterministic in mathematical form, but they assign only probabilities to the location of observable particles. William Stoeger says, 'In the projection of this quantum world into the macroscopic world of our experience, determinism fails' ('Epistemological and Ontological Issues Arising from Quantum Theory', in *Quantum Mechanics*, p. 89). That is to say, when one of the many possible futures described by the wave function is 'chosen' by making a measurement, it is objectively uncertain which future will be so chosen. For quantum mechanics, the same cause can give rise to different effects. This was a great shock for those who thought that science must assume as a basic axiom that the same cause always produces the same effect. It is, however, welcome to believers in radical freedom, whose view of freedom is that the same initial situation can open up a number of different tracks into the future.

For believers in radical freedom, a paradigm of a morally free decision is when one possible action realises an objectively good state, and another realises the egoistic satisfaction of the agent. Believers in radical freedom, among whom I count myself, think that the ascription of praise and blame to humans depends upon agents being undetermined by anything other than themselves, their knowledge of possibilities open to them, their awareness of right and wrong, and their capacity to choose without compulsion in crucial moments of moral choice.

Determinism is the view that every physical state is the only possible effect of preceding physical states plus the laws of nature. It depends on the axiom that the same cause always produces the same effect. If this axiom is undermined by quantum physics, the postulate of radical freedom becomes much more plausible.

2. THE PRINCIPLE OF INDETERMINACY

Some theorists maintain that radical freedom can operate in the indeterministic gap opened up by the collapse of the wave function

(William Pollard, *Chance and Providence*, London, Faber, 1958). The mind can choose which possibility to actualise, and that is its area of freedom. But I agree with John Polkinghorne that this seems an unduly 'hole-and-corner way of influencing quantum events' (John Polkinghorne, *Science and Providence*, SPCK, London, 1989, p. 28). Quantum indeterminacies even out over large numbers, so that by the time we get to the level of ordinary experience the processes of nature are governed by quasi-deterministic Newtonian laws (as has been seen, they are not truly deterministic). We might be unwise to confine human freedom to the realm of quantum indeterminacy.

The importance of quantum theory is not that it proves that radical freedom really exists. It might seem much too restrictive to make morally responsible free acts depend on tiny random events in the sub-atomic realm, which mostly cancel each other out. The principle of indeterminacy does not even strictly prove the world is not deterministic. It certainly sets limits to our ability to predict the future. But the world might be determined, even though we are unable to predict it. There are quantum theorists, from Einstein to David Bohm, who advocate a deterministic substratum for quantum events. Bohm has proposed a deterministic account that envisages a completely different approach to quantum theory, re-establishing particles as a central feature of the unobserved world, but seeing particles in a completely non-Newtonian way, such that their properties cannot be specified independently of a holistic context (that is, a context in which the nature of the parts is partly determined by the structure of the whole). This is determinism of a distinctly non-atomistic and non-mechanistic kind. Whether it is truly deterministic at all is open to question.

It cannot be said that quantum theory has wholly established objective indeterminism, even though most standard interpretations of the theory (especially those deriving from the Copenhagen interpretation) are indeterministic. James Cushing, Professor of Physics and Philosophy at Notre Dame University, remarks that 'considerations of empirical adequacy and logical consistency alone do not force one to accept an indeterministic view of fundamental physical phenomena' ('Determinism and Indeterminism in Quantum Mechanics', in *Quantum Mechanics*, p. 99). The issue, Cushing holds, is probably not empirically decidable.

However, given the success and wide acceptance of a Copenhagen-type interpretation, what quantum theory does show is that full physical determinism is not necessary to good scientific procedure and explanation. Indeterminism within specifiable limits of probability does not undermine scientific explanations. Since the majority interpretation of quantum physics is actually indeterministic, the fact is that much available scientific evidence counts against determinism.

For those who are inclined anyway to believe in radical freedom, such freedom entails a lack of complete physical determinism in the universe. What quantum theory shows, minimally, is that good and fruitful science does not entail such complete determinism. More positively, in what is still its standard, though disputed, interpretation, quantum theory suggests that indeterminism is a basic feature of the physical universe.

This in turn suggests, though it does not entail, that 'even at those macroscopic levels where classical physics gives an adequate account, there is an openness to the future which relaxes the unrelenting grip of mechanical determinism' (Polkinghorne, *Science and Providence*, p. 30). The point is that, once the stern grip of determinism has been loosened at the quantum level, we might be more open to the possibility of a macroscopic universe that is not bound by determinism. Copenhagen-type interpretations of quantum theory do not prove that the macrocosmic universe is not physically determined. But they open up the possibility that it might not be so determined, and that macrocosmic indeterminism, the necessary condition of radical freedom, is wholly compatible with modern science.

3. INDETERMINISM AND FREEDOM

There is thus no scientific barrier to indeterminism at the everyday level. No scientific laws will be broken if humans can on some occasions determine by an act of will, not itself scientifically observable, that one future rather than another should become actual. Physical determinism is a dogma of philosophy, not an axiom of science.

This thought is reinforced by the fact that there is no hope of proving that, *on every occasion*, one and only one future outcome of a state of affairs is possible. It is, after all, an extremely ambitious statement to say that there is only one possible future at every stage of the existence of the universe. How could we possibly establish that no other track into the future was even possible? As the philosopher David Hume put it, what strange sort of necessity would it be that ensured that every causal situation could have one and only one outcome?

It does not look as if we could ever establish such a thing scientifically. For a start, we know that we cannot exhaustively specify every causal feature in any real life situation. We do not have the resources to measure all real life situations in a rigorous experimental way. As I shall show in the next section, quantum theory shows that events in the universe are entangled, so that events light years away, and unknown to us, can produce changes of state in phenomena that we are observing. That entails that we cannot have exhaustive knowledge of any local physical state without having exhaustive knowledge of the whole universe, which is impossible (see John Polkinghorne, *The Quantum World*, Longman, London, 1984, chapter 7).

Quantum physics provides another reason why it is impossible to gather exhaustively all possible information about any given state. Heisenberg's uncertainty principle prevents us being able to measure precisely both the position and momentum of a fundamental particle, and so it prohibits complete knowledge of any observed physical state. Thus we cannot know the precise causal conditions of an observed state of affairs at any stage of the universe's existence. That means that we can never be in a position to say precisely what effect is bound to follow from such causal conditions.

The consequence is that we can never conclusively test whether physical determinism is true – whether one and only one precise effect is bound to occur, given any causal state and the laws of physics. That is because we can never establish exactly and exhaustively what any specific causal state is and therefore we can never predict exactly what effect is bound to follow. That does not mean physical determinism is false. But it does mean that its truth cannot be scientifically established.

Physical indeterminism – the view that on at least some occasions, however rare, more than one possible future can follow from a given physical state of affairs, and no physical factor determines the actual future in advance – is thus fully consistent with present science, it is accepted by most quantum physicists and it cannot be scientifically disproved. That is a fairly strong position for a theory that common sense and morality presuppose when they assume that at least some physical organisms – namely, human persons – are sometimes free to choose which track into the future they will take.

Whatever the truth or otherwise of determinism, it is certain that many causal outcomes can only be predicted in a probabilistic or statistical way. This is true, not only at the tiny quantum level, but also at the level of ordinary life. The development of chaos theory shows that dynamic systems far from equilibrium have the property that minute fluctuations can, in the right circumstances, have large-scale effects. We have all heard the story of how the flap of a butterfly's wing can cause a hurricane on the other side of the world. The point is that we cannot predict many ordinary states (like the weather) if they depend on fluctuations in minute sub-atomic conditions. At this point chaos theory connects with quantum theory, because quantum theory shows that there are such tiny fluctuations, which are unpredictable in detail (that is true, whether you believe in 'ontological indeterminacy' or not). If such unpredictable fluctuations give rise to macrocosmic changes, it follows that such changes will be unpredictable too. Therefore deterministic prediction, of the sort Laplace hoped for, is in principle impossible: 'Chaos brings a new challenge to the reductionist view that a system can be understood by breaking it down and studying each piece' (J.P. Crutchfield et al., 'Chaos', *Scientific American*, 225(6), 1986, p. 56). There are significant patterns in the way the world is, but they permit an openness of texture and an access to novelty that 'provides a mechanism that allows for free will within a world governed by deterministic laws' (ibid.).

All this does not prove that freedom and creative spontaneity are objective features of the universe. What it shows is that determinism cannot be proved, that it is not needed for good science and that many scientists these days do not accept it. The standard interpretations of quantum theory posit that the cosmos changes in accordance with causal principles that are stochastic or probabilistic. To many modern scientists, it looks as if the universe at the quantum level continually moves into an open future by probabilistic processes. It is a plausible generalisation from this that similar processes may occur at the atomic and molecular levels. Suitably complex organisms may give rise to acts of creative spontaneity and responsible freedom in a world that lays down a number of alternative tracks into an open and undetermined future. That is a very different picture from that of the heirs of Newton and classical, largely deterministic, physics.

4. THE QUANTUM WORLD

It must be stressed that this whole area of physics is still a realm of deep disputes. Einstein never accepted that a statistical theory could be complete, and so maintained to the last that 'hidden variables' could one day be found that would eliminate probability. The physical world, he believed, did consist of objective events that could in principle be modelled by a complete physical theory that would be deterministic and would not speak of such strange objective realities as probability waves, superimposed realities, collapses of wave functions and non-physical interactions between fundamental particles.

This does not entail the sort of 'universal physical determinism' that was inspired, however oddly, by Newtonian mechanics. It affirms that there are real physical particles and real non-probabilistic causal laws, and so restores a generally Newtonian view of the physical world. But it need not deny, any more than Newton did, that there are also non-physical causal factors in reality with which the physical sciences do not deal.

In any case such a theory is not within sight and even the most ardent Einsteinian would have to agree that the hope to find a 'realist' view of fundamental particles that are more or less ordinary physical entities is a matter of faith. God may not play dice with the universe, but it *looks* as if God does!

David Bohm's attempt to develop a theory that would restore determinism to fundamental particles is at the cost of postulating a mysterious 'pilot wave', without energy, probably unobservable, conveying information faster than light. His precise theory has not convinced many physicists, but it shows that quantum physics does not necessarily rule determinism out. It also shows that physicists are nowhere near having a complete knowledge of the bases of matter. Physicists are as yet unable even to reconcile relativity and quantum theory in one coherent explanatory theory. Perhaps determinism is not, after all, dead. But it is absolutely not established by the sciences and remains little more than a philosophical dogma.

Furthermore it looks as though any adequate account of both determinism and matter will be very different from any simple picturing theory of intuitively clear realities. The quantum world is one of virtual particles continually fluctuating in and out of existence, even in a vacuum; of superstrings vibrating and twisting in ten or eleven dimensions, most of which are 'rolled up' and invisible; of a reality in which gravitation, space-time and matter are or were merged in one 'instanton' (a postulate of Hawking and Turok). This is not anything like the classical materialist world of small massy particles bumping into each other for ever.

Determinist or indeterminist, the reality with which modern physics deals is one in which solid things seem to disappear into probability waves, in which space-time can be finite, unbounded and expanding without expanding in anything, and in which new space-times can possibly be generated inside black holes. This is a world far beyond common sense and anything we can picture to ourselves in a simple way. So modern physics shows the world to be a reality that is very different from anything the senses perceive and to be one that is deeply intelligible and mathematically beautiful.

In a sense modern science has, in a strange and unexpected way, returned to a form of Platonism. An important part of Platonism says the senses provide only appearances, whereas reality is purely intelligible and knowable, if at all, only by mind. Platonism, it seems, has returned in triumph, now guided and tested by observation and measurement, but erected on a foundation of intellectual beauty, perceptible only by mind, a world of concepts and numbers appearing to us as a world of material objects constructed by our senses.

This may not be a religious view, but it carries unmistakeable resonances of the basic religious insight that the whole sensory world is an appearance of a deeper spiritual reality. That reality is one of supreme beauty and wisdom and is knowable only by a disciplining of the mind and will. In this world of appearance, human freedom has chosen tracks into the future that have realised suffering and frustration as well as moments of creative insight and altruism. The religious path is one of seeking to re-orient a freedom that has preferred sensory desire back towards the primal beauty from which it has fallen.

Naturally, quantum physics is not concerned with such issues. But, for those who are, the picture suggested by at least some quantum physicists of a probabilistic cosmos, founded on a hidden reality of beauty and wisdom and yet making possible moral freedom and creativity, is one that is deeply consonant with a religious view.

8

THE IMPLICATE ORDER

The quantum view of the universe is very different from classical materialism. It sees the universe as a holistic system, the parts of which are interconnected in intricate ways, very different from the world of apparently distinct solid objects we see and touch. The universe can be seen as an emergent whole, advancing through creative novelty to ever more complex levels of self-organisation and information processing and storage. Quantum theory is neutral as to the existence of God, but if there is a God this is the context in which human lives need to be seen and in which the purposes of God are to be worked out. It is a grand and inspiring context, which calls for a rethinking of many traditional religious beliefs. But it is not at all in conflict with belief in the existence of a God of supreme wisdom and a concern for intellectual beauty and creative diversity.

AIM: To show that the cosmos should be seen as an organic whole, a system whose parts are modified by the nature of the whole system. If God is the ultimate environment of the cosmos, that within which the cosmos 'lives and moves and has its being' (Acts 17:28, quoted from the poet Epimenides), then God's nature and purposes will affect the nature and action of its fundamental particles and basic physical forces.

SUGGESTIONS FROM QUANTUM THEORY:
C. INTERCONNECTEDNESS

1. ENTANGLEMENT

A major feature of quantum theory which seems to be well established but is also astonishing is quantum entanglement. It seems that change is not always brought about by objects bumping into one another in space, but can be effected instantaneously in non-local ways (that is, between entities widely separated in space). There are non-local, superluminal (faster than light) connections between particles that have once interacted. The Bell theorem, constructed by CERN (Conseil Européen pour la Recherche Nucléaire) physicist John Bell in 1964, states that elementary particles are 'entangled', that once two particles have interacted with one another they remain linked, even when they have moved vast distances apart. The universe seems to be a vast network of interacting particles, bound into a single quantum system. 'In some sense the entire Universe can be regarded as a single quantum system' (Paul Davies and John Gribbin, *The Matter Myth*, Penguin, Harmondsworth, 1992, p. 217).

It has been known since 1864, when Maxwell invented field theory, that the waves that constitute physical reality are infinite in extent. Gravity extends as far as space extends, though it is so weak at great distances that its effects are not discernible. Quantum theory goes further. Not only are its waves probability waves, so that they are only physical in a very strange sense; they are also connected in what seems to be a non-physical way, since nothing can travel faster than light, yet particles instantaneously correlate with one another over vast distances.

Things in the universe are connected to other things in the universe in a non-local way, yet a way that produces real and observable physical effects. So Bohr says, 'Isolated material particles are abstractions' (Herbert, *Quantum Reality*, p. 161), and David Bohm has developed the idea of the universe as an 'implicate order' in which 'particles may be strongly connected even when they are far apart, and this arises in a way which implies that the whole cannot be reduced to an analysis in terms of its constituent parts' (David Bohm and B.J. Hiley, *The Undivided Universe*, Routledge, London, 1993, p. 6).

Bohm supports the idea of holistic explanation, in which the behaviour of the parts of a system can only be explained in terms of the whole system of which they are parts. This means that the behaviour of things in the cosmos is not to be explained just in terms of their smallest physical parts, as in old-fashioned atomism. It is to be explained, ultimately, in terms of the whole cosmic system. For a theist, that whole cosmic system is itself grounded in a deeper all-enveloping reality, the mind and being of God. If there is ever going to be an 'explanation of everything', it will have to involve an understanding of the mind of God. That, of course, places it beyond the bounds of natural science, which is very annoying for some physicists.

Even if we do not have such grandiose ambitions as an explanation of everything, the fact of quantum entanglement shows that we cannot consider events in the universe in isolation. We have to learn to see them as parts of wider systems that help to determine their character.

Einstein wrote, 'A human being ... experiences himself as something separated from the rest – a kind of optical delusion of his consciousness' (Herbert, *Quantum Reality*, p. 250). We seem to be independent beings, but in fact we are parts of an interconnected unitary universe. Our sense of separate consciousness is, says Einstein, 'a delusion'. The word 'delusion' is rather strong, but it could be misleading to think of ourselves as individuals who could exist in just the same way without this physical universe. In this phenomenal world, we seem to be separate independent beings, but in reality all beings are strongly interconnected or entangled. That does not mean, as Einstein's words may suggest, that we are not autonomous individuals, with our own distinctive sense of self. But it does mean that we are not as easily disconnected from our environment as we might think. We are internally, not just contingently, connected to it after all.

If we could see reality as it is, we would see that we could not have come to exist in any universe other than this, and that even small changes in our small part of the universe may have effects at immense distances and in quite unforeseen ways. It may seem as though we could just make some changes that would make this world better – remove pain, for instance. But it may be that the changes this would produce in the complex wave-form that is the universe would be absolutely devastating. The possibilities for changing the universe radically may be much narrower than we imagine.

The dream that sometimes occurs to each of us, that we, just as we are, could have existed in a much better universe, could never become real. We may fantasise about living in a universe where everyone is caring, kind and compassionate. What we are apt to forget is that if everyone was caring, kind and compassionate, we would be excluded from that universe, since we would not be caring, kind and compassionate enough to gain entry. For better and worse, we belong just where we are.

The prospects may look bad for immortality. And so they probably are unless, having been born and grown up in this universe, personal beings can take a new form of existence in a different sort of universe altogether. Interconnectedness within this cosmos would then be an essential feature of the origin and development of such personal beings. But personal beings might be capable of being transferred to a quite different form of reality. That is a dream of many religions. Strangely, science does have something to say about it. But that comes later, in part 3.

2. THE MIND OF GOD

The quantum world is extremely strange, and it may be that some new theory awaits that will change our view of basic physical reality in an unforeseeable way. But it is vastly unlikely that physics will ever return to the sort of simple materialism that some (wrongly) thought to be an implication of Newtonian mechanics. Reality is not the world it appears to be to common-sense observation. It is a world in which every part is connected to every other, so that discrete entities are slices out of overlapping and dynamic fields of potentiality. Particles are what appear to us, with our form of consciousness. Complete knowledge of the underlying reality is forever hidden from us, yet the structures we can know have an elegance and beauty that is astonishing.

In the face of this theory, I think we have to renounce the possibility of complete understanding of physical reality and yet celebrate its immense intelligibility and complexity. A veiled, interconnected, probabilistic and yet intelligible reality underlies the physical reality in which we live and move and have our being. Quantum theory is, of course, neutral as to the existence of God. There are quantum theorists who deny the existence of any personal creator of the universe. But all would accept that the ultimate nature of physical reality is not the world of discrete and solid atoms circling in the absolute space and time of Newtonian mechanics. It is a literally unimaginable world of elegant beauty, which is more mathematical or intelligible than material (i.e. solid and precisely located in space). It is perhaps a quasi-Platonic world, exhibiting a complex interplay of chance, intelligibility and necessity. We discern a minute part of its phenomenal appearance as its internal dynamic structures of possibility creatively actualise into the diverse observed forms of space-time and into us.

It is a natural, though not forced, step from here to postulate that, just as the world of phenomenal appearances exists in and for finite minds, so the hidden reality that underlies it exists in the infinite mind of God. This is a natural move, for it is plausible to think that mathematical or intelligible realities, which are largely possibilities or probabilities, can only truly exist objectively insofar as they are conceived by a mind that is fully actual. So we can think of one ultimate mind that conceives the non-material basis of the physical cosmos. Materialism is turned on its head.

The physical cannot even exist without the spiritual, without a consciousness that conceives and actualises it.

That consciousness might well be called God. Such a God would not be a person external to the universe, interfering with it from time to time. It would be more like the mind of the universe itself, the ultimate reality within which all things, intelligible and physical, exist, and the infinite reality part of which all finite things express. But this may seem unlike the God of traditional religion – Paul Davies has contrasted the 'God of the physicists' with the God of religion, to the advantage of the former (Paul Davies, *God and the New Physics*, Penguin, Harmondsworth, 1983, p. 229).

We might say that quantum physics, though by no means committed to the existence of God, can suggest an idea of God as supreme intelligence and cosmic wisdom. It can give rise to an idea of God as a consciousness conceiving the 'veiled reality' of mathematical forms and possibilities (suggested by the primacy of consciousness); as the creative lure into the openness of an undetermined future (suggested by indeterminacy); and as the totality of mind that structures the ever more complex organisation of a developing cosmos (suggested by interconnectedness). But it still tends to be uncomfortable with what is widely seen as the religious God, the fully personal, active and miracle-working God. That is a gap I will try to close in part 3.

3. THE SCIENTIFIC WORLDVIEW – NEW VERSION

The progress of physics since the seventeenth century has been astonishing. It has changed our view of reality irreversibly. That progress began with a rejection of the Platonic view that the intelligible world is more real than the phenomenal world, and with an anti-Platonic insistence that the real is the temporal and particular and that knowledge is to be obtained by observation and experiment, not by pure theory. But in the twentieth century there was a revival of a sort of Platonism. For many quantum physicists, the world in itself is not observable or even imaginable, and is the real and intelligible basis of an observed reality that is only an appearance to human consciousness. Yet this is a different sort of Platonism. Observation and experiment are essential, and all pure mathematical theories must in the end be subject to confirmation or disconfirmation by controlled observation, even though such confirmation may be very indirect.

According to the Copenhagen interpretation, the world of objects-appearing-to-consciousness, although it is a world of appearances, is nevertheless fully real. It adds to the reality of the intelligible world that underlies it, which after all is only a world of potentialities and probabilities. The phenomenal world cannot be reduced to unreality or unimportance by the discovery that there is a deeper reality underlying it (as, arguably, it was for Plato or at least for some later Platonist philosophers), for it actualises that veiled reality in creative and distinctive ways.

One interpretation of quantum physics is that the basic reality is conceptual, intelligible and not exhaustively knowable by us. Phenomenal reality, the everyday world known by us, is partly constituted by finite consciousness, and its laws are selective generalisations that cannot completely capture the partly indeterminate richness of reality, and do not picture reality as it is apart from our observation of it. As William Stoeger writes, 'The laws of nature as we actually formulate them in the sciences are but imperfect and incomplete descriptions of the regularities ... that obtain in reality itself' ('Epistemological and Ontological Issues Arising from Quantum Theory', in *Quantum Mechanics*, p. 95).

If we attempt to put the Galilean, Newtonian, Darwinian and quantum theoretical innovations together, to provide one coherent scientific worldview, we are likely to get a picture of the cosmos as an interconnected, emergent and organic unity. Human consciousness is an integral part of this universe, not an intrusion of the spiritual into the material, and not the only goal for the sake of which the whole universe exists. Finite consciousness is a product, not of a deterministic and mechanistic clockwork universe, but of a universe in which both creative spontaneity and holistic forms of causality play an important part. That is to say, the causes that operate in this universe are often free and spontaneous projections into a partly open future. The way they operate depends upon the nature of the whole cosmic system and the part they play within it. They may, as Arthur Peacocke suggests, operate in a 'top-down' or 'whole-part' way, the system as a whole affecting in the most basic way the behaviour of its parts (Peacocke, *Paths from Science towards God*, pp. 51–56, 108–14). So we can only fully understand the causal processes of the universe when we understand the nature of the goal to which the universe moves and the nature of the total system that constrains the activity of its parts. For a theist, that total system will include God, as the highest and most inclusive system of all.

Thus seen, the cosmos is a complex interplay of chance, freedom, intelligibility and necessity. The observed universe is a phenomenal appearance of an underlying intelligible realm of potentialities, which are continually being actualised in new ways, unpredictable in detail yet constrained by the finely tuned values of the universe's basic constants and forces and by the past interactions of its entangled properties. As the cosmos grows more integrated and complex, it generates beings that can understand and to some extent shape their own future, and this trajectory of growth points towards the genesis of self-aware and self-directing minds as the ideal outcome of the universe's long journey of self-realisation.

4. ALTERNATIVE WORLDVIEWS

So what is the view of the cosmos given to us by science? Is it a pointless, impersonal, blind and purposeless shuffling of atoms, arising by chance from chaos, driven by forces of cruel competition and ruthless destruction, doomed to final annihilation? Or is it a cosmos of profound intellectual beauty, exquisitely fine-tuned to manifest intrinsically worthwhile states of intelligibility and awesome magnificence, generating communities of finite subjects possessing creative freedom, the capacity of comprehending the cosmos, and unique forms of affective experience, the fragility and finitude of which only increase the preciousness of their transient value?

I doubt if a decision can be made on purely scientific grounds. Many other background factors, partly dependent on personality, personal experience and perspective, are at work. From a purely scientific point of view it is hard to make a decision. Modern science has undoubtedly been used to construct a sophisticated atheistic worldview. One popular view of the progress of modern science is that God has been progressively driven out of the picture until the whole idea of God has become superfluous, an unnecessary complication we can well do without. However, I have aimed to show that the story is more complicated than that.

The Galileo case has been seen as a battle between science and religious faith. It does show that most traditional religions, like most ancient philosophies, held mistaken views about the nature of the physical universe. It also shows the repressive tendencies that any dominant ideology, whether religious or secular, can exhibit. But the case is best seen as a struggle between competing scientific theories. Its main implications for religion are to query very literalistic interpretations of Scripture and to suggest that human beings are probably not the only or even the main concern of God. These, I think, are wholly positive suggestions, which religion can accept with thanks.

Newton's work has been used to support a materialist and determinist worldview, in which nature consists of purely material particles obeying absolute and unbreakable laws. But this was not Newton's own view, and it does not at all follow from his discovery that there are general laws of interaction between those physical entities whose nature can be specified quantitatively. The materialist worldview is not even implied by the hard science. The idea that the universe usually operates according to laws, not according to the arbitrary acts of spirits, is an important insight, but it is entirely in accordance with the belief that God is a rational creator of an intelligible universe. Again, then, we could see the contribution of science to religious faith as a wholly positive one.

Many atheists have seen Darwin as the champion of their cause, though Darwin denied ever having been an atheist. However, the theory of 'progress' though random mutation and natural selection is a powerful explanation of how complexity and seeming design could emerge by purely natural processes, without particular acts of divine interference. Darwinism does make a purely physicalist account of the universe much more plausible than it has ever been. But it does not *entail* a physicalist account, and Erasmus Darwin's view of evolution as generally directed towards a goal by a wise creator remains highly plausible to many. Some biologists find the organised complexity of organic life so astonishing that for them the postulate of an intelligent designer of the laws of nature is overwhelmingly probable. If that were so, then again science would contribute positively to religious insights by seeing God as creating a universe in which creatures gradually emerge that are able partly to shape their own futures, and in which the physical naturally evolves into the personal.

If Darwin has become the champion of atheism, Heisenberg and the quantum physicists have placed a huge question mark over the materialist worldview. That has not, of course, put an end to atheism. Some see in quantum mechanics and in Einstein's relativity theory the possibility of giving a complete account of the universe in some relatively simple general theory, a 'theory of everything'. Such a theory may, for suitably cautious physicists, only mean that all known laws of physics might be incorporated in one general theory, which is not an impossible goal. But it is often taken in a stronger sense to mean a theory that also explains just why the laws of nature are as they are, and how consciousness can be explained in purely physical terms. If this stronger version could be achieved, God would be unnecessary. There would be nothing left for God to do, since everything could be explained in terms just of a few fundamental laws. However, there seems little prospect that such a theory could actually be achieved. It is, at best, a dream of science, not an established hypothesis. And there are deep puzzles about where or how the fundamental laws could exist without any physical universe, which might make us sceptical about whether such explanations could ever be truly ultimate.

Indeed for many quantum physicists something like God becomes necessary to the very existence of the material universe. For observation and consciousness may well be fundamental for the existence of material entities. The whole of material reality becomes a realm of appearance, and the 'veiled reality' that underlies it may include consciousness as an essential element. That would reverse the materialist worldview entirely. It would re-instate a generally Idealist view of the world, reminiscent of that of philosophers like Leibniz, Kant, Hegel and Berkeley. It would make God, not superfluous to science, but necessary to the existence of any material reality. There could hardly be a more positive contribution to belief in God than this.

Interpretations of quantum physics are too varied and preliminary to enable any strong conclusions to be drawn. But it should be clear that the story of modern science is not a story of the progress from theism to atheism. It is now, and it always has been, a story of increasing insights into the nature of physical reality, which naturally change some religious views, just as they change previous scientific views, but which can as easily form a positive contribution to religious belief as they can undermine it.

It is true that the universe as seen by modern science does not support the idea of a benevolent disembodied person who is able to do absolutely anything and who ensures that all sentient beings are protected from harm and will live for ever in the best of all possible worlds. That is one way in which modern science might change some religious views of God – though such views were always rather naïve and obviously contradicted by everyday human experience.

On the other hand, it has become increasingly clear to most scientists that the fundamental laws of the universe are elegant and subtly complex. They have produced the immensely improbable existence of seemingly free and rational personal beings with the capacity to understand and to act with purpose for goals deemed to be worthwhile and desirable. Even if the universe has no purpose or value, it has immense intelligibility and has given rise to beings that do have purposes and values. In that sense, the universe can well seem orientated towards, directed towards, conscious life. It is thus plausible to see it as an intelligible reality coming to awareness of its own intelligibility, through a process of developing self-organisation and, in its most recent stages, of purposive striving.

5. THE MIND OF THE COSMOS

Questions remain about the basis of this intelligibility, about whether this process is truly purposive and about whether conscious awareness and understanding are an intelligible goal for the universe. We may remain agnostic about such questions, since they do not seem susceptible to any scientific answer in the foreseeable future. But if we think that the universe does seem to show direction or purpose, we may consider that only a conscious and rational being can formulate a purpose (an envisaged and desired future) and cause it to be pursued in an efficient manner. So the idea of a conscious being of enormous rationality that causes the cosmos to exist for the sake of the intellectual beauty it exhibits, and so that the cosmos may achieve the goal of developing self-organisation and self-understanding, is one that is naturally suggested, but not entailed, by the scientific worldview.

Such a being would not be personal in the sense of being whimsical, capricious or prejudiced, as human persons are. But it would be personal in having some analogy to what we understand as reason, intention and awareness. The aspects of nature that we are unable to explain in terms of the actions of a purely benevolent supernatural person may yet be seen as inevitable in an intelligible and goal-oriented cosmos of this sort. A self-shaping cosmos entails a degree of open possibility, and therefore of possible mistakes and failures, rather than a securely determined progress to a completely pre-specified goal. The nature of sentience in a law-governed emergent universe probably entails suffering as well as pleasure for organisms living under general laws that cannot always ensure their freedom from harm. The degree of that suffering can be enormously intensified by the unwise or malicious choices of finite agents who could and should have chosen otherwise. In this way we could see the universe arising from a cosmic mind of great wisdom and good purpose, constrained by necessities inherent in the sort of universe this is, in which alone sentient carbon-based life-forms like humans can come to be.

In this part I have tried to outline the fourfold revolution that the sciences have brought about in human thought. That revolution has on the one hand encouraged the view that God is superfluous to scientific explanation and that the methods and conclusions of science are very different from those of traditional religion and philosophy. However, I have suggested that the scientific revolution can alternatively, and plausibly, be seen as supporting new and important insights into belief in God as the ultimate mind of the cosmos. It is no accident that most of the key scientific figures I have mentioned were either wholly committed to the existence of God, or at least did not see any implication of atheism in their work.

It is true, nevertheless, that the scientific idea of God that has emerged is sometimes thought to be very different from a religious idea of God. The God of science is a super-intelligent consciousness who shows a concern with intelligibility and intellectual beauty in creation, but who has little if any personal interaction with creatures and seems to show little interest in their individual well-being. This God functions largely as an ultimate explanation of the universe, especially of its rational structure and incredibly organised complexity.

But is such an ultimate explanation really necessary? And is it possible for science alone to provide a truly ultimate explanation of why the universe is as it is? In part 2 I will argue that there are a number of very important aspects of reality that are beyond the reach of scientific explanation. These will have to be taken into account in any 'theory of everything'. If so, it follows that the final theory of everything cannot be purely scientific. Science alone is incapable of providing such a theory, though it will be essential in providing part of it. What more is required, and how can it be provided? That is the subject of part 2, which will argue that to provide the sort of ultimate explanation for which science longs it is necessary to look more widely than science, to examine the limits of a purely scientific approach, and to explore what might be said about reality that is beyond the province of the natural sciences.

PART II

THE SEARCH FOR ULTIMATE EXPLANATION – THE GOD OF THE SCIENTISTS

9

THE LIMITS OF SCIENCE

*The development of the modern scientific worldview has set a new
context for believing in God. God must be seen as the basis of a vast
cosmic process of elegant mathematical beauty, emergent complexity and creative self-organisation. Humans are a small part of this
process. Science has transformed our image of God. Yet there are
important aspects of reality that lie beyond the self-imposed limits of
science. Science deals with the publicly observable, measurable
and predictable. Beyond the reach of science lies the realm of value,
purpose and personal experience. This realm must be incorporated
into any adequate account of the nature of reality. It must be part of
any ultimate explanation of things, which therefore must be more
than scientific.*

*AIM: To show that the natural sciences do not deal with
ideas of value, purpose and significance. Yet such ideas may be
important for determining the nature of ultimate reality. This is
especially so if there is a God, a reality of supreme value and
cosmic purpose.*

1. THE LIMITS OF OBSERVATION

Modern science begins with the ejection of purpose, value and significance from the universe. This is one main reason why the 'scientific worldview' fails to deal with all aspects of reality. The 'disenchantment of nature', the stripping away of all personal properties from the mechanisms of nature, was important to the birth of modern science. Aristotelian science included final causality as one of the explanatory features of nature. All things have an ideal end at which they aim, even if the ideal of being a tiger is only to be a good member of the tiger species. But the search for final causes proved fruitless, and experimental science only flourished when it excluded such a search and instead sought laws of relationship without purpose between objects.

The exclusion of the personal from nature is a methodological axiom of science – at least until the advent of some interpretations of quantum theory. With it goes the exclusion of any consideration of value in nature. For the idea of value is essentially connected with the idea of personal reality – a 'value' is something that is valued by some conscious personal being, whether God or some finite being. Such a distinctively personal reality came to be called in question when it was held that all knowledge was to be justified only by observation and experiment.

When it is said that data are observable, it is meant that they are publicly observable – anyone in the right conditions can observe them, whatever their emotional state or character. When it is said that they are testable by experiment, it is meant that such observations can be repeated in experimental situations in a predictable way. What enables you to make such predictions is the fact that sequences of events fall under precise laws of nature. For instance, the law of gravity tells you that two bodies will attract one another proportionally to the product of their masses and inversely proportionally to the square of the distance between them. For this to be so, you have to be able to measure mass, position, distance and velocity exactly. If you do, you can then predict exactly how the bodies will behave relative to one another (if no other force intervenes). This prediction can be confirmed repeatedly by duplicating the initial situation, and everyone can see it for themselves. That is one ideal model of a scientific explanation.

But does this show us what reality is? It shows that close observation and exact measurement reveal that physical objects interact in regular, predictable ways, ways that we seem to be able to capture quite neatly in the form of a few relatively simple general laws. This is quite astonishing, and very few people before the sixteenth century would have thought it was true. Isaac Newton's formulation of the laws of mechanics did indeed introduce a new way of looking at the physical universe, as a law-governed set of objects and properties in space-time.

So there is little question that the laws of science do show us many important things about reality. This is not a universe in which many spirits or gods, good and bad, battle for the souls of human beings, causing disease, plague or odd moments of good fortune. It is not a universe in which a benevolent God is always doing whatever is best, or in which human acts always receive their just deserts. Here again scientific findings undermine any view that the way things go is always directly willed by some god or spirit. This is a universe in which material objects interact in accordance with general laws that seem indifferent to particular moral or immoral acts, or to the desires of capricious spirits and demons. For experimental science, it does not seem that the universe is morally ordered in particular ways. Its laws rule over good and bad alike, and cause events to occur without apparent reference to the beneficial or harmful effects they have on human or other sentient beings.

2. CRITERIA OF TRUTH

That is an important discovery, but it does not answer the question 'And is that what reality is?' For if that is what reality is, you have to believe that the observational sciences give us the only access to truth. You have to believe that the only good reason for believing something is that you have observational, experimental evidence for it. But once you put it like that, it is obvious that such a hypothesis is self-refuting. The belief that you should only believe things on the basis of observation and experiment is not itself based on observation and experiment, and therefore it is stating that it should not itself be believed. Such a statement of belief is as near to a wholly self-refuting statement as anyone could desire. You are saying, 'Only accept as true what can be observationally established.' But no observations show that statement is true. So it seems that you should not accept it after all (though you might accept it as a recommendation that you are free to reject if you can think of some other way of establishing truths).

I think what lies behind this recommendation is the thought that to accept only what the sciences can establish is the most reasonable or economical course, and it works, producing useful, fruitful knowledge. If so, we are accepting the more general principle that rationality and fruitfulness are basic tests for truth, and what we shall then look for is the most reasonable and fruitful account of the nature of reality. Part of a reasonable account is that it should cover all the different sorts of data there are in as coherent a way as possible. And part of a fruitful account is that it should be helpful in maintaining a morally admirable, socially helpful and personally integrating way of life. Of course, any such account must be consistent with the best scientific findings. But can we be quite sure that only scientific procedures are rational and fruitful? Might there not be other data, not accessible by the methods of the natural sciences, with which our account of reality needs to be consistent as well?

3. DREAMS

It is not hard to think of data, perfectly familiar to all of us, that are not covered by a strictly scientific account of reality. One typical example is dreams. Dreams do exist, for virtually everyone has them. If I frame the sentence 'There are dreams' (if I quantify over dreams, to use the technical jargon), it makes perfectly good sense. There is observational evidence that nearly everyone dreams. In psychological tests in Edinburgh University, people are woken up when they display rapid eye movement, and asked if they are dreaming. They always say 'yes', even though, if they were to wake up naturally much later on, they may deny that they have dreamed at all. So there is experimental evidence that we all dream, though we usually forget our dreams.

If you do forget them, there is no way in which anyone else can know what you were dreaming about. Moreover, if you do say you dreamed about something, there is no way in which anyone else can check whether what you say is true or not. Your dreams are not publicly observable or repeatable. They do not follow any laws of nature and they are not predictable using such laws. But surely you did dream something; there was something there (wherever 'there' is, in the case of dreams).

Now if I accept the recommendation 'Only accept as true what can be observationally established,' I would never believe what you told me about your dreams. You observed your dream, no doubt. But I have to just trust you as to what the dream was about. I have no access to it. Even your own observation cannot establish that what you say is true, because you only have your memory to rely on, and we know how unreliable that is.

There is an easy escape from this dilemma. We can just say that we accept many reports of personal experience (like dreams) as true, because we trust the person who has such experiences, not because we can measure these experiences, or put them under general laws, or repeat them or observe them. Here, it seems, is an entirely natural case where we accept that something is probably true, even though it is beyond the reach of scientific method. Dreams are not physical objects in space, so observational science cannot get a grip on them.

4. THE REALITY OF PERSONAL EXPERIENCE

It could be argued that one day dreams will be correlated with brain states, and we will then be able to explain them scientifically, by examining the dreaming brain, and knowing what the brain is thinking of because we will have a sort of dictionary that tells us what images various brain states correlate with.

But how would we set up a dictionary for correlating brain states with the thoughts and images that occur in dreams? We would have to ask the dreamer what he or she was experiencing when the brain was in a specific state. Having done that, it might just be possible to tell what the dreamer was experiencing by some sort of brain-scanning (as long as there were no dreams that had never been dreamed before, for which our dictionary would have no entries). But to get the information in the first place, or to confirm that our guesses were correct on any subsequent occasion, we would just have to ask, and trust the answer. Whatever the neurological sciences may establish about dreams, it is impossible to escape the conclusion that at some stage we have to accept data that are not given by public observation. We can then try to incorporate them into a scientific theory, but the point is made: public observation, with which the natural sciences deal, is not the only means of access to truth-claims about reality. There are forms of human knowledge and experience that fall outside the reach of observational science.

There are reasons (perhaps drawn from science itself) to think that there are forms of reality with which observational science cannot deal. As well as dreams, there are also mental images of all sorts, thoughts that we never tell anyone about, and all sorts of pains, itches and feelings that we may keep to ourselves. I am more certain I experience such things than I am of any philosophical theory that tells me they do not exist.

5. THE UNIQUENESS OF EXPERIENCE

It is not just a matter of itches, dreams and feelings. Even when I observe the public world, my particular way of observing it is unique and virtually inexpressible. Think of hearing a particular chord in the middle of a Beethoven symphony. The playing of the chord is a public event, and the orchestra is publicly observable.

But when I hear the chord, I do so in the light of (I should say 'in the sound of', but there is no such expression in English) the music that has gone before. Perhaps it is the resolution of a discord, or the completion of a melody. Perhaps I am hearing the piece for the first time and so cannot quite make sense of it. Or maybe the piece is very familiar, and I am comparing it with previous performances, as to timbre, speed and resonance. I will also be anticipating what comes next, and that too will depend on how well I know the piece and on how I think it might be interpreted.

Involved in this apparently simple experience of hearing a musical chord are my own musical skills, my memories, anticipations and emotional reactions. All these intertwine in a complex way, so that even that momentary experience is quite unlike the experience of anyone else who is hearing the same chord at the same time. I could not put it into words, but it may be a very important event in my life, of great intensity and significance – or it may not. Nobody will know but me – unless I tell them, very imperfectly and clumsily. Even then, perhaps nobody will know quite what it is like for me to have that experience.

Such experiences include some of the most important things in any human life. Standing before the Grand Canyon at sunset, catching a glimpse of someone you love, being suddenly transfixed by the beauty of a Rembrandt portrait or catching a wave at Big Sur – these can be the most vivid and meaningful experiences in a human life. To deny their existence is absurd, and to dismiss them as nothing more than a series of electrochemical events in the brain is to prefer a speculative and unconfirmed theory to the testimony of direct experience.

All such personal experiences are beyond the reach of the natural sciences. There are no measurable quantities, no general laws, no equations to capture their quality and intensity. They do have quality and intensity – they are pleasant or painful, weak or strong. This very fact shows that they are real, not imaginary. But there is no precise scale against which they can be measured. We might identify a part of the brain that is active when we feel pleasure, measure the intensity of electrical discharge and correlate it with our feelings of emotional strength. But this can only be done in a crude and general way. For if we cannot even properly describe our experience of hearing a symphony, what exactly are we correlating with brain activity? It could only be experiences that we can describe in a very inadequate way, except to say that they are generally pleasant, intense, moving or exhilarating. Human experiences are intrinsically fuzzy, opaque to description and hugely complex. They do not break down into atomic units that obey a few simple laws of interaction. They are essentially interconnected – present connects with past and future in ways that have never before existed and that will never be repeated exactly.

6. THE ABSTRACTIONS OF SCIENTIFIC THEORY

The situation is quite different for the physical world as it is described by science. In atomic theory, for example, every particle of the same sort (e.g. an electron) is identical in basic qualities such as mass and electric charge to every other. The laws that govern their interactions are simple and universal. Scientists explain events by reducing their components to simple elements and bringing these under general laws. But if I have a new thought, it is unlike any other thought that has ever existed (if it is really new, which is perhaps rather rare, though it must happen). It does not arise by the application of any repeatable law. I cannot find any atomic constituents of thought that are all identical, or any general laws that can produce original ideas.

It is the uniqueness and interconnectedness of experiences that make scientific methods inappropriate for dealing with them. We can speak of intense, long-lasting, pleasant experiences, we can devise methods likely to produce certain sorts of experience and we can vaguely sort experiences into general categories. So there can be no serious doubt that experiences exist and have quality, intensity and duration. But they do not have the repeatability, discreteness and denumerability of physical objects.

This helps to bring out, by contrast, how fortunate we are that the world of physical objects is not like that. For most of human history, people did not realise that physical objects are reducible to identical sorts of constituents that are exactly measurable and obey more or less universal and relatively simple laws. The possibility of science, in other words, is not at all obvious or necessary. There might have been a physical world in which, like the world of experiences, every object was unique and not precisely measurable, and changes occurred in accordance with principles that were only general tendencies and patterns rather than exact laws.

Then again, the suspicion may arise that maybe the physical world is really like that. Quantum theory suggests that it may be much more like that than classical physicists imagined. Could it be that the scientific worldview, perhaps the greatest change there has ever been in human understanding, works precisely because it abstracts from the unique particularity of events, artificially isolates elements that are intrinsically interconnected, and imposes a grid of rigid laws on to a physical reality that is in fact more plastic and flexible?

I think that is probably the case. Nonetheless, the physical world is publicly observable, measurable and subject to general laws to a sufficient extent that modern science is possible – whereas this is not true of personal experiences.

There are personal experiences, known to all of us in a direct and natural way, that do not fall within the domain of the natural sciences. The scientific domain is that of publicly observable objects in shared public space. Since science does not deal with personal experiences, it cannot itself give an account of what they are or of how they relate to objects in physical space. Science itself cannot provide a comprehensive worldview, because there are aspects of reality with which it does not deal. The most obvious aspects of this sort are personal experiences. It is precisely in such experiences that such notions as value and purpose have their home.

7. PERSONS AND VALUES

An important feature of personal experiences is that they are usually value laden. Many experiences are perceived as carrying a specific value with them – so pain is evaluated as bad and pleasure as good; some sights are evaluated as beautiful, others as ugly; some tastes as pleasant and others as unpleasant. There may be many types of evaluation, of various degrees of importance. But it is rare for an apprehension to occur that is not given some value by its experient.

The importance of this feature is that it is entirely lacking in the natural sciences. Events occur in accordance with general laws, and that is neither good nor bad, pleasant nor unpleasant. Physical facts, as observed by science, carry no values. A scientist may feel awe and amazement at the complexity and elegance of natural processes. But that should be classified as a subjective reaction to observation, not as any sort of objective property of physical objects. Or we may say that the scientist's personal experience of the physical order may be highly value laden, though the physical process itself is value neutral. In contrast, my view of the Grand Canyon at sunset is intrinsically awe inspiring. The awesomeness is a property of my experience. The scene is not just a value-neutral succession of events, to which I may have a specific reaction. Not to feel awe is to miss something that is there to be apprehended.

People have different experiences of the Grand Canyon, but that is precisely the point. They have different experiences of the same physical array. They do not have the same experience, to which they adopt different subjective reactions. This is, in a way, just to say that experiences *are* subjective. But that is a point of great importance, for to call them subjective is to say that there are realities that are not publicly shared and that carry values with them. In other words, values are experienced as realities, though as 'subjective' (not publicly shareable) realities.

8. VALUE, PURPOSE AND SIGNIFICANCE

With the idea of 'value' go the ideas of 'purpose' and 'signifi-cance'. For a purpose is the goal of a process or activity. It is what such a process aims at, or is directed to attaining. To have a purpose is to aim at a state considered to be of value, and a process is purposive only if it aims at a valued state.

An event is significant, has meaning, if it can be seen as playing an important part in the realisation of some purpose or goal, or if it is itself intrinsically worthwhile. Without the notion of value, the notion of purpose would not make sense. The basic reason that the physical sciences do not deal with purposes is that they do not deal with values. They deal with a value-neutral reality, and so no natural process can be seen as purposive or properly goal directed.

Scientists may speak of the 'purpose' of the heart being to pump blood around the body and to keep the body alive. This is necessary to the body flourishing, and if that is seen as good it might be helpful shorthand to speak of purposes. But any biologist will point out that the heart is not really trying to keep the body alive. The alleged purpose is in fact just a consequence – the body lives because the heart works. There is no real objective purpose there. To get a real purpose you would have to say that someone thinks the existence of the body is a good thing.

Of course we do think the existence of bodies is good. So we can see the beating heart as purposive, from our point of view. But again this is a subjective evaluation. Natural science of itself allows no purposes. To do so, it would have to conceive of a being who valued or intended the existence of certain states – something very like a God. Believers in God may reasonably claim that God sees bodies to be good, since God wants persons to exist, and so there really are objective purposes. But the sciences do not mention or presuppose this, for the basic reason that God also falls outside scientific investigation, and so is ignored by natural science.

9. PURPOSE AND THE COSMOS

It is obvious that scientists cannot measure or observe God, or bring the acts of God under some general testable law so that they can be precisely predicted. Unless we are going to rule out the possibility of God by definition, this is another area beyond the limits of science proper. It carries the consequence that if God ever acts in particular ways in the world, those acts will be beyond the scope of science to explain, and the physical events that are the consequences of divine acts must have causes at least one important component of which science cannot measure or explain.

This, it should be emphasised, is not a problem for theism and does not bring science and religion into conflict. It simply demonstrates that science cannot explain everything, or hope to bring everything under purely physical explanatory laws, if there is indeed a God who acts. But why should it? Is it not enough for science to demonstrate the elements and laws that make up the physical realm, without going on to claim that there is nothing else and that everything in the physical realm must fall under a set of purely physical laws and not arise in any other way? That is just a dogmatic claim that no spiritual reality can exist. And it is a claim that scientific practice cannot prove, since if one thing is clear in modern science it is that we are far from knowing everything about the physical basis of nature. We cannot give an exhaustive account of its causes, and it is probable that we will never be able to confirm that we know all the causes of every event and know that there are no others.

The realm of science ignores value, purpose and significance. It ignores consciousness and personal experience. It ignores any purely spiritual reality such as God. Science brackets out such things, takes no account of them. This is a decision of principle, and it has been immensely fruitful. The reason for this is not hard to find. If there are real objective purposes in the cosmos, one obvious place to locate them is in the mind of God. But the mind of God is inaccessible to human minds. Just as we cannot know what an author intends for the characters in a novel, so we cannot know what God intends for this universe and the beings in it. In a novel, however, we can gradually come to build up a picture of what is likely to happen – though there may always be surprises. So in the cosmos we can guess at some of the divine purposes, and gain a general impression of their character, from inspection of the history of the cosmos. This will be a matter of seeing what sort of purposes there could be in the universe if there is a God, not a matter of proving there is a God by establishing that there definitely are purposes in the universe.

10. THE LIMITS OF SCIENCE

What scientists deal with is the measurable, predictable and regular operation of objects, as such objects exercise their natural powers in interaction with other objects. It is an astonishing fact that there are comprehensible laws detailing such things. Their existence is a necessary condition of our being able to understand and eventually to control natural forces. What science cannot do is prove that no other sorts of reality exist, or prove that physical objects only ever act in the predictable and regular ways with which science deals.

There are, then, very real limits to science. This is not a matter of things science cannot yet do but might one day do. It is a matter of the limits science imposes upon itself, in confining itself to public observation, repeatability, law-like regularity and measurability. One extreme form of the scientific worldview is the belief that this is the only sort of knowledge there is and the only sort of reality there is. But that could not be a scientific statement, since it is meta-scientific, a statement about what science is and deals with.

Perhaps there are other sorts of reality than the public and physical, and perhaps even the public and physical contains supra-scientific elements. Most religious views do take that alternative view. In doing so, they do not conflict with science. They conflict with reductive materialism, with the belief that nothing exists except matter. But it was always obvious that they would. If there is a conflict, it is about whether science has any limits and whether it gives all possible information about physical reality. This is not itself a scientific issue. And that is the proof that there are questions of fact, of what the nature of reality is, that are not scientific questions.

10

THE EXPLANATION OF EVERYTHING

Any truly ultimate explanation of the universe will have to include considerations of value, purpose and personal experience. In view of this, I propose that the most adequate ultimate explanation will be in terms of an actual being, an ultimate mind, that necessarily contains in itself all possible worlds and states, and that actualises some of them for the sake of the distinctive forms of goodness they realise. God, as the ultimate explanation of the universe, unites physical explanation – the predictable regularities of nature – with personal explanation – the intrinsic values for the sake of which such regularities exist. Worlds can contain evil as long as this is eventually positively related to good, of which the evil is a necessary condition or consequence, and the realised values are overwhelmingly great. In the case of ultimate mind itself, those values are likely to be supreme and unmixed with evil, so that ultimate mind is also a reality of supreme value.

AIM: To show that the metaphysical idea of an objective God, an ultimate mind of supreme value, is a natural extension of the scientific quest for ultimate explanation. That is a good reason for thinking God exists – natural theology is not dead! To show that a true consilience of science and culture must unite and preserve the notions of value and laws of nature.

1. THE PRIORITY OF MIND

One of the basic decisions about worldviews is whether you decide to start with consciousness and experience as the fundamental clue to your interpretation of reality, or whether you look for a wholly physical account of reality, thinking of consciousness as probably an accidental by-product of physical processes. Most religious believers are committed to making consciousness primary, for the simple reason that they think of the ultimate reality as a form of conscious reality that is in most cases considered to be the source of all physical reality. Certainly for a theist all material reality depends upon a spiritual or conscious reality, which is therefore the more fundamental of the two.

Science as such is not committed to denying this. But, since science deals only with the physical, since it seems to have established the general dependence of human consciousness on matter, and since matter existed long before any finite consciousness existed, science has sometimes been used in support of a materialist worldview, for which mind is at best totally dependent upon matter, or at worst an illusion.

There is an alternative hypothesis. Perhaps mind or consciousness is the fundamental reality, and the material is dependent upon mind. This alternative was taken for granted by Galileo and Newton. It was accepted by Charles Darwin, even though he came to doubt particular acts of divine providence and the benevolence of the creator. And it is accepted by many modern physicists – Paul Davies and Nick Herbert are just two examples – though they do not always associate the underlying mentality of the cosmos with any religious idea of an interested and interacting God. The God of the scientists is real, but tends to be a rather impersonal, Platonic God, a beautiful but austere mind unconcerned with the problems of human life, expressing its being in the elegance and intelligibility of the universe, yet indifferent to its cruelty.

Why should the idea of such a God appeal to so many scientists? It is because the existence of brute matter, which just happens to be the way it is for no reason and yet gives rise to such a fine-tuned, utterly improbable and beautiful universe, terminates the quest for understanding in a way that is repugnant to any scientist. For scientists are committed to press the question of understanding as far as they can. They can accept that there are limits to the methodology of natural science, but find it hard to accept that there is no further way of pressing towards a deeper and final understanding of existence. Could the postulate of a cosmic mind help to further that quest?

If you think such a mind would simply be a concatenation of thoughts, feelings and intentions, its existence would be just as puzzling as that of the material universe, and just as unsatisfactory as an ultimate explanation of existence. But that is not the sort of mind scientists are interested in. They look for an ultimate intelligibility, something like a final self-explanatory mathematical theory that cannot be other than it is and whose necessity in some way generates this material universe. The search for a 'theory of everything', for a deep necessity underlying all observed realities, is one expression of that urge to understand that drives science. But the strictly scientific search for a theory of everything seems doomed. For it seems odd to think of mathematical theories existing on their own, even before any physical realities exist. It is rather less odd to conceive of a God in whose mind mathematical theories could exist. But mathematical theories do not seem to be systems that exclude all alternatives, since there are typically many alternative mathematical axiomatic systems, not just one finally explanatory one. Moreover, and most seriously, there seems to be no way of 'breathing fire into the equations', of accounting for the physical existence of a contingent cosmos from purely mathematical facts.

What is needed for an ultimate explanation of everything, if such a thing is possible, is something that transcends scientific method, that is not just a set of abstract mathematical axioms, but is actual in a full sense, and that can generate physical reality. That would be an actuality richer and more stable and enduring than that of physical reality, intelligible in itself and necessary in its nature, yet capable of creativity and understanding.

2. THE NECESSITY OF MIND

One way to conceive of such an actuality is to recall the hypothesis, widely held among mathematicians, that mathematical truths are fully existent only when they are conceived by some consciousness – that is, they are ultimately products of the mind. We can then say that mathematical truths can exist even before the existence of the physical universe, if they exist as products of a supreme mind. So we can frame the idea of a consciousness in which all mathematical structures exist. That consciousness could select one axiomatic system and construct a physical universe that was patterned on that system. The hypothesis of such a consciousness exactly fits the bill for something that might be an ultimate explanation of the universe.

Generalising from this mathematical example, we can posit that this ultimate consciousness might be able to conceive of every possible state of affairs – every possible world – that could ever exist. We can posit that there necessarily exists a complete array of every possibility of any kind, something like the Platonic world of Forms. This array is necessary, for every possibility is exhaustively expressed within it. No other possibilities exist, and the possibilities that do exist are necessarily what they are. No question arises of why this consciousness is as it is, since it includes all possible worlds and states exhaustively. No reason needs to be given why one state exists rather than another, since all these states exist, though only as possibilities. But it is plausible to think that possibles can only exist if there is some actuality that sustains them in being. That actuality is mind, which conceives them, and is necessarily what it is, the actual being that is necessary to give these possible worlds real existence as unactualised possibilities.

This form of argument is just the one used by classical philosophers like Plato, Thomas Aquinas and Immanuel Kant when they were seeking something that could be an ultimate explanation of the world. A variant of the argument suggests that there need not be a set of all possible states, complete and ready for inspection in the divine mind. Rather, that mind may creatively imagine new possibilities that do not pre-exist in any sense. Such possibilities must be based on something already existent in the divine mind if they are not to be simply arbitrary. So in some sense the patterns or archetypes of the general sorts of possible states that can exist must be in the divine mind, even if there is also much scope for the radically free creation of 'new' possibilities. This view allows for radical novelty even in conceiving what may possibly exist. I am sympathetic to this idea. It makes the dependence of possible states on a divine intelligence and imagination that can conceive of them even more important.

Although these arguments are very abstract, they will be familiar to modern cosmologists, who formulate theories that postulate many, perhaps an infinite number of, possible worlds and try to conceive necessary mathematical structures in their quest for an ultimate explanation of why things are as they are. We can hardly claim to understand such ultimate factors completely, but we can say that such ideas seem to be coherent; and, if they are, we can see how they might provide an ultimate explanation of reality, even though we have no access to anything like an immediate intuition of such an explanation. God, the supreme self-existent mind, can be an ultimate explanation of the universe and yet remain a mystery to human understanding. We dimly see what such an explanation might be, but we, being limited in our powers of comprehension, can never completely grasp it.

3. ULTIMATE EXPLANATION

The explanation would go something like this: ultimate mind is the actual basis of all possible states. It is the only being that must be actual, if anything at all is possible. It is thus uniquely self-existent, not deriving its existence from any other being. Its nature is necessarily what it is – there are no possible alternatives to it, since it is the basis of every possibility. It can be spoken of as omniscient, in the sense that it conceives or generates all possible states, knows what they are and knows that there are no more than it conceives. It can be spoken of as omnipotent, in the sense that it brings whatever is actual into existence from the realm of possibility, or it generates actual beings with a derivative power to make some possible states actual. Nothing that comes into being can have more power than ultimate mind has, since the latter is the one ultimate source of all actuality.

It might be as well to note that these definitions of omniscience and omnipotence are not exactly the same as the ones classical philosophers have often given. Many philosophers define omniscience as knowledge of absolutely everything, possible, actual, past, present and future. They define omnipotence as the power to do absolutely anything that is not self-contradictory. The definitions I have given are more restricted than that. They do not entail that God knows what will be actual in the future. Perhaps God leaves the future open for radical freedom. And they do not entail that God can do absolutely anything. Perhaps God leaves, or even must leave, finite reality to follow its own inherent laws of development.

Yet we can still say that God knows everything that is possible and actual (the future may not be actual yet) and that God is the most powerful being there could possibly be and the ultimate source of all things that come into being. This leaves open the question of exactly which possible states can be made actual and whether there are restrictions on what possible states can be actual. Though such an ultimate mind can sensibly be called omniscient and omnipotent, this may not be enough to satisfy some religious believers. It is enough, however, to satisfy the requirements of being an ultimate explanation of the universe.

To complete such an explanation, there needs to be some reason why some possible states should be selected for actuality. I think this can be provided, if we think that some possible states are of greater value or desirability than others. Agents may desire many different things and may disagree about what things are truly desirable, but it seems obvious that, in general, some states are more desirable than others. Imagine God trying to decide between creating a universe in which all conscious beings suffer terrible pain and frustration for ever, and another universe in which all conscious being are happy, wise and loving. It is obvious that the second universe is better, more desirable, than the first.

Choices between possible worlds will be much more complex than this. Because of factors outlined in part 1 such as entanglement, the generality of physical laws, and the moral freedom of finite persons, many valuable states will only be able to exist in a universe in which there are also many undesirable states, involving suffering and frustration. Whether it is good to create such universes is a question only God could answer. But there will be some universes God could easily create, some very bad universes God would never create and a great many universes that will be mixtures of good and bad. Between them we might think of God as exercising a sort of spontaneous, creative, choice. Then we could say that the reason a universe exists is because it is a possible universe that God makes actual for the sake of the distinctive sorts of goodness it contains. The creative choice of ultimate mind is a good explanation for why particular universes exist.

The philosopher Leibniz thought that a rational God would have to create the best of all possible worlds, the best possible set of states that could ever exist. But I agree with Thomas Aquinas in doubting whether there is just one 'best possible' world. It seems more likely that there are many possible worlds that are good in different respects, but that none of them is absolutely the best. So it seems to me that it is enough that God creates a set of states that exhibit great and unique sorts of value, even if the world they are in is not absolutely the best.

In any case, actualising possible states because they are truly valuable or desirable seems to provide the best possible reason why, out of all the possible states there are, some would become actual. And that would be the best possible explanation of the existence of an actual universe.

This does not mean that ultimate mind would have to create a physical universe with other conscious beings in it. It could actualise many desirable states just for itself, and contemplate and enjoy them. As Aristotle said, it could find the highest happiness in contemplating its own perfection – its own perfection including all the desirable states that it actualises in itself (*Metaphysics* 12:6). This will include indefinitely many forms of beauty, intelligibility and bliss. The process may be likened to the activity of a sort of divine imagination, conceiving possible states, selecting some of them for their distinctive value, actualising them and finding pleasure in contemplating them.

This may all seem highly speculative, and it is important to remember that what we are doing is asking what, so far as we can understand it, an ultimate explanation of the universe might be. This is a properly scientific question. But any attempt to give an answer takes us well beyond the limits of the natural sciences. For it takes us into the realm of mind, of consciousness, reasons, goals and intentions, a realm to which the physical sciences have only limited access.

My suggestion, which follows philosophers like Aristotle and Leibniz, is that the best ultimate explanation of the universe is an actual being that contains in itself, and is aware of, all possible states and that chooses to actualise some of them because it finds them to be of intrinsic value – that is, they are valued just for their own sake.

If it actualises such states in its own being, the self-existent divine mind may be rightly called the Supreme Good, since it will choose for itself a set of highly valued states. That set is not in an entangled cosmos, containing elements of chance and the free acts of other conscious beings. So God will be without imperfection or frustration and can ensure that the goodness of the divine being is very great, and greater than that of any other being it actualises. The divine mind will be, as Aristotle put it, supremely happy in the contemplation of its own supreme perfection. Aristotle, not so far as we know any sort of religious believer, called it God.

4. THE MULTIVERSE

Isaac Newton called the space and time of this universe the 'sensoria of God', logical spaces in which beautiful and intelligible objects of direct divine knowledge are given actuality, moving from the realm of the possible into the objective existence of the actual, and then being received back into the divine mind through knowledge of them as actual.

It could be that, for some reason unknown to us, this is the only possible universe that could exist. That would certainly answer the question 'Why does this universe exist, rather than some other, different, universe?' Some physicists do try to show that there is only one finally consistent mathematical system that could give rise to a physical universe. That would remove creative choice from God, but God would still be the conceiver of possible worlds, the giver of physical actuality, and the Supreme Good.

Most cosmologists, however, suppose that there are a great many possible worlds other than this one, worlds that could well have existed and may even actually exist, though we could never apprehend them directly. How many such possible worlds are there, and how many of them may become actual? We have seen that in modern cosmology the idea of a 'multiverse' is often thought to be the most plausible explanation of the fine tuning of this universe. When Martin Rees, the British Astronomer Royal, in his book *Our Cosmic Habitat* (Princeton University Press, Princeton, 2001), considers the amazing fine tuning of our universe, he says, 'We seem to have three choices: we can dismiss it as happenstance, we can acclaim it as the workings of providence, or (my preference) we can conjecture that our universe is a specially favored domain in a still vaster multiverse' (p. 162).

A multiverse is a vast – but how vast? – ensemble of universes, of space-times, that all exist. Perhaps, some physicists suggest, black holes can spawn other space-times, with differing initial conditions and laws of physics. Perhaps, as Everett suggests, all possible states specified by a quantum wave function in Hilbert space actually exist and our universe is just one series of such states. We could never communicate with such other universes. But if there were a vast range of them, perhaps even every possible set of laws and initial conditions, then the existence of this fine-tuned universe would no longer be a surprise. It would be bound to happen sooner or later.

Appeal to a multiverse, within which our universe is just one case, still leaves massive problems unsolved. I am sympathetic to the idea that there is an exhaustive array of all possible laws and conditions from which particular existent universes might arise. This is just a variant of the classical theistic hypothesis that God knows every possibility – though a theist might add that perhaps the creative mind of God can create infinitely new possibilities. There may be an infinite number of possible states. In that case, there could never be an exhaustive set of all actual universes, since at every stage, however large, there would remain an infinite set of unrealised possibilities.

So how do some possibilities get chosen for actualisation? Many cosmologists adopt the axiom that 'even the most unlikely events must take place somewhere' (Max Tegmark, 'Parallel Universes', *Scientific American*, May 2003, p. 41). That is, to put it bluntly, false. There is no a priori principle that guarantees that every possible state will come into existence. And that is a very good thing. It is extremely unlikely that I should boil my mother, cut her up into small pieces and serve her up at a dinner party. Must there be a universe in which I actually do that?

What prevents me eating my mother is that it is wrong. Just because it is possible that I could be evil, does it follow that, in some universe, I actually am evil? The most repugnant aspect of some versions of the multiverse theory is that, according to them, there must exist universes that are totally morally repugnant, in which sentient beings are totally irrational and in which they all suffer unending and excruciating torments for no reason.

Tegmark thinks (as Martin Rees apparently does) it is a benefit of multiverse theory that the strange fine tuning and other coincidences of this universe can be wholly explained. If every possible universe exists, such coincidences must exist in some universe, so they are not strange after all. The consequence would be, however, that nothing at all would be too strange, too irrational or too immoral to exist. Any sane human mind must draw back in horror from such a thought. In such a system, there would not just be one Hell, which would be bad enough. There would be an endless series of Hells, each bad in its own way.

On the God hypothesis, however, many possible universes are precluded from existence precisely because they are too irrational or evil. The amount of suffering in this universe is certainly great enough to cause intellectual difficulties, yet the universe contains enough intelligibility, beauty and happiness to make it a candidate for actuality. If we could show, or even plausibly suggest, that suffering was ineliminable from this universe, and that it could somehow be made to relate positively to otherwise unobtainable forms of goodness, then we could account for the actualisation of this universe in terms of a selection of possible values by an ultimate mind.

This is a point at which science's wilful neglect of value leads to fantastic and horrifying consequences. On Tegmark's view, not only must every possible universe, however evil, exist, but it must be repeated an infinite number of times. If possibles exist only in the mind of God, however, then only those possibilities will be actualised that exhibit some preponderance of value. Each actual universe will retain its own unique distinctive value. And the fine-tuning coincidences will be what they appear to be, evidence of the selection of cosmic parameters for the sake of generating living intelligent beings.

5. THE PRINCIPLE OF PLENITUDE

Although it is beset with difficulties, the idea of a multiverse prompts a further suggestion about the nature of the primordial creative mind. If we posit a divine mind, then an infinite and exhaustive array of possible universes could exist in the actuality of that mind. Perhaps an essential characteristic of primordial mind is the principle of plenitude. According to this principle, it is good that every possible sort of good should exist, as long as its existence does not come at the price of excessive and pointless harm.

St Augustine comes near to framing the principle when, discussing the existence of evil, he writes that the world's course is 'more gracious by antithetic figures' (*City of God* 11:18). That is, 'As a picture shows well though it have black colours in divers places, so the universe is most fair, for all those stains of sins' (11:23). Augustine's idea is that the universe is better if there are many grades of goodness, even if the lower grades involve the existence of evil. From this we might generalise that the more kinds and degrees of goodness there are the better. Perhaps there is something in the divine nature that causes it to generate many possible sorts of good, even though some kinds will inevitably incur suffering.

We would be wise to add a proviso that the good must overwhelmingly outweigh the evil and that evil must in some way be transformable. That is, any evil must be capable of being placed within a wider context in which it will be part of a complex good, or it must be capable of giving rise to an otherwise unobtainable good. An example of the former would be the way in which suffering entailed in pushing the body to its physical limits actually becomes part of the complex good of doing well in a race. Or we might come to see how our suffering is a necessary part of a universe that is of overwhelming goodness. An example of the latter would be a case in which the patient endurance of your own suffering makes you much more sympathetic to the suffering of others.

The suggestion is emphatically not that it is actually worth positively creating such suffering in order to get those goods. It is rather that the creation of a world in which disciplined and compassionate persons exist may carry with it the actuality of some suffering and the possibility of much more. The suggestion is that the creation of such a world by a being who creates for the sake of goodness is only justifiable if the suffering that arises – either necessarily or as a result of finite acts in opposition to the desire of the creator – can meet the condition that the suffering can be transformed by being related to some positive good.

Further, it is probably the case that this good must be experienced, at least to some degree, by the sufferers themselves, so that finite subjects who suffer do not just become means to the end of the happiness of others. Whether the world as we experience it meets this condition is questionable. Much suffering seems to be completely without point, in the lives of innocent victims. That is a major reason for theists to think that there must be some form of life beyond earthly death, in which the possibility of transformation can exist.

The principle of plenitude would lead you to expect a plurality of worlds, in many of which there would be a mixture of good and evil. In many worlds, it would not be possible for most finite persons to see how the evil was to be positively related to future good. But, given the possibility of continued existence in a successor world, such a positive relation could be realised.

The human condition does seem to be one in which many quite distinctive values are realised, though it also carries with it a great deal of suffering and frustration. If there is a successor world to this in which sentient beings can continue to exist and the evil they had experienced could be positively related to good – a life after death – it is perfectly intelligible to suppose that, out of all the array of possible universes that exist in the mind of God, this one contains sufficiently unique and distinctive kinds of good to make its actualisation desirable and worthwhile.

Thomas Aquinas too asserts something like the principle of plenitude: 'A mark of active will is that a person so far as he can shares with others the good he possesses.' So 'it befits the divine nature that others also should partake of it' (*Summa Theologiae* 1a, 19:2, responsio). Aquinas does not draw from this the conclusion that may seem naturally to follow, that it necessarily belongs to divine perfection that there should be many created goods, and many subjects to experience them, and perhaps that the more different kinds of goods there are the better. It seems logical to do so. Or at least it could be held that there is some form of necessity in the divine nature that leads it to create certain forms of finite being that can actualise and appreciate goodness. And it may be that many forms of finite goodness necessarily involve the actuality of some evil and the possibility of more. It would in my view be a complete explanation of the existence of suffering in the universe if it could be shown that the possibility of much suffering and the actuality of some suffering is necessarily implied by the structure of each member of a class of universes (e.g. universes containing free intelligent agents) some of which necessarily exist.

Some philosophers take a very short route through these intractable problems, and just say that, since God is omnipotent, God could create a universe without any suffering at all. Such a view supposes that humans can understand the ultimate nature of necessity, of what ultimate reality must be and of its inner nature. The Scottish philosopher David Hume held at one point that whatever we can imagine without self-contradiction must be possible. 'It is an established maxim in metaphysics', he says, 'that nothing we imagine is absolutely impossible' (David Hume, *A Treatise of Human Nature*, Part 2, Section 2, J.M. Dent, 1974, p. 39). So if we can imagine God creating a world without suffering, God could do it. The subsequent progress of mathematics has proved this view to be mistaken. There are many mathematical proofs that things humans (think they) can imagine are impossible and that things are possible that humans cannot imagine. We can imagine, for instance, that there is a highest prime number (a number only divisible by itself and 1); but we can prove there is not. We can imagine that arithmetic is in principle a complete and consistent system, but Godel proved that it is not. We can imagine that there is some square equal in area to a given circle. But we can prove that there is not. We can prove that electrons are probability waves in Hilbert space, but we cannot imagine that (envisage it in a mental picture).

So modern mathematics shows that human imagination is not a good guide to the nature of ultimate reality. Hume was wrong. He was nearer the mark when he asked, 'Who can retain such confidence in this frail faculty of reason as to pay any regard to its determinations in points so sublime?' (*Dialogues Concerning Natural Religion*, Part 1, Harper, New York, 1972, p. 6). There may be necessities in things that we cannot comprehend. When we speculate about ultimate explanations, the best we can do is think up some axioms, trace out their consequences as well as we can, try to see whether they fit the observed facts, and just hope that we are not completely mistaken.

In this spirit, I will concede that Aquinas is probably right in supposing that, from the fact that it is good to create as many different sorts of good as possible, it does not follow that God must create every possible sort of good. All that follows is that, if God creates any world containing a very large set of valuable states, even if that world contains many ineliminable evils, such creation is a good thing, subject to the conditions set out below. So we cannot know how many universes God actualises. We cannot say that God must actualise as many good universes as possible. We cannot say that God must actualise the best possible universe, if there even is such a thing. We cannot be sure that God must actualise some universe like this one. All we can say is that, if and insofar as God is supreme goodness, any universe God does actualise must contain or make possible a very large and distinctive set of values. All evil in it must be capable of being understood as a necessary condition or consequence of the existence of those values, and must be transformable in some way so that it is not pointless and final. Given those conditions, we can intelligibly say that an adequate reason for the creation of a specific universe is that it realises great, distinctive and otherwise unobtainable values.

6. NECESSITY AND CREATIVITY

Augustine thought of the possibility that there might be 'worlds without end', that is an infinite number of different universes (*City of God* 12:19), though he was reluctant to make a decision on the issue. Where ancient saints hesitated, modern cosmologists rush in to assert the existence of many universes. But when they posit a multiverse, they are not, as Martin Rees implies in the quotation above, positing an alternative to God. They are suggesting that the divine mind produces not just one universe, but many, and perhaps an indefinitely large number of, universes. If God is indeed perfect beauty and bliss, it seems plausible that

God could allow any possible universes to exist that meet the conditions of the principle of plenitude. There will be many possible universes that God would not permit to exist, since they would contain too much untransformed evil. But it may be in the nature of God, as self-realising pure actuality, to allow into actuality, to 'let be' (Genesis 1:3), many possible worlds that are overwhelmingly good and in which evil and suffering are necessary conditions or consequences of great goods that sentient beings enjoy in those worlds.

God would then be seen to be not only perfect beauty and bliss, but also the infinite creator of unlimited kinds and degrees of goodness, through the overflowing divine plenitude of perfect being. The corollary would be that God would have to be able to transform evil in any world that comes to be.

In this scheme both necessity and contingency, changeless perfection and creative freedom, spirit and matter, exist and are coherently related. The array of all possible states and the actuality of that in which they exist are necessary. But the actualisation of specific sets of possibles is contingent. Such actualisation does not follow from the description of possible states alone, but, while concurring with the principle of plenitude, allows free creativity in the specific worlds that are brought to be. The divine perfection is changeless, since it can never cease to be the most perfect possible being. Yet the specific content of the actualised objects of divine contemplation is generated by creative freedom. Spirit is the fundamental reality, without which there could be no possible states and no material states. But the material world provides at least part of the content of divine consciousness, new and relatively autonomous objects for divine contemplation. It is in such a coherent combination of necessity and freedom that an adequate ultimate explanation of the universe will lie.

7. THE THEORY OF EVERYTHING

The cosmos, on this view, is not just a self-existent or accidental reality. It is an expression of the divine mind, chosen for the sake of the distinctive values it makes possible, by a fundamental reality that is in itself the Supreme Good. Such a conclusion could never be part of natural science, for it speaks of that which is beyond the natural – the personal origin of the cosmos. But it is a conclusion that shares the most profound scientific motivation, which is to understand why the cosmos is as it is. It advances human understanding of the cosmos in a way that, though hesitant and imperfect, makes comprehensible the beauty and intelligibility of the physical world.

This cosmos is actualised by mind from the exhaustive array of possibilities in the divine consciousness, and the cosmos exhibits both the rational necessity of intelligible law and the free creativity that arises from its own spontaneous projections into an open future.

If there is a theory of everything, it must lie in the mind of God. The mind of God is hidden from human understanding. Nevertheless, the human mind is capable of seeing that the natural and reasonable completion of its quest for understanding lies in the existence of a divine self-existent mind. It is in this sense that modern science can be seen as pointing beyond itself towards religion, or at least to belief in ultimate mind that is in itself the perfectly good and beautiful. Science can correct the myopia that often afflicts religion when religion is only concerned with appealing to the supernatural to help obtain human desires. Yet religion can point to the completion of the insistent search for understanding that drives the natural sciences. It can point to this completion in something that lies beyond and yet perfects science, the mysterious yet in itself wholly intelligible reality of ultimate mind.

8. TWO FORMS OF EXPLANATION

Natural science is the search for understanding and explanation. But more than one type of basic explanation is needed if human beings are truly to understand the universe in an adequate way. Some philosophically minded scientists seem reluctant to accept this fact. In his book *Consilience* (Little, Brown, London, 1998), Edward O. Wilson says bluntly, 'There is intrinsically only one class of explanation' (p. 297). What is this favoured class? Wilson says, 'All tangible phenomena ... are based on material processes that are ultimately reducible ... to the laws of physics.' The ultimate and only sort of explanation is that found in physics.

It is highly dubious whether the whole of biology and chemistry can be reduced to physics, let alone disciplines like psychology and sociology. The real problem, however, is that persons undoubtedly exist as parts of the natural order. When persons are considered, as they are in the social sciences or in history, some things that happen are explained not by saying that they inevitably follow by some general principle from a preceding state, but by saying that they realise a desire or goal that is more or less intelligible. Such a form of personal explanation in terms of goals and intentions is commonplace, and different in kind from causal law explanation. It is different in kind because it uses quite different concepts. Physics makes no mention of desires, goals, intentions or in general of mental states. That vocabulary simply is not part of physics.

The necessity with which physical explanations deal is the necessity of physical change in accordance with universal principles that are usually not themselves necessary. Such explanations do not deal in preferences or values at all. Preferences and values are rooted in desires, which are internally related to choices of states of value, states that a rational agent has a good reason to choose.

It is possible to state in a formal way what an intrinsic value would be – a state chosen by a conscious personal agent, just for its own sake and not for the sake of some further end. There must be some intrinsic values, some things that people choose just for their own sake, if there are any values at all.

Personal explanations try both to set out what intrinsic values there are, and the values that are instrumental to obtaining them, and to explain personal conduct in terms of attempts to discover what they are or to attain them. All personal explanations terminate in the specification of states that are thought by someone to be of intrinsic value. They explain sequences of human acts in terms of an intention to realise some such values. The reason such statements could never be translated into laws of physics is that neither intentions nor values are so much as mentioned in physics.

The forms of physical and personal explanation are quite different. In the former, we have an algorithmic process of applying rules to transform physical states. The outcome is simply what results from the application of the algorithm; there is no sense in which the result is 'foreseen' by the physical process itself at its beginning. In the latter, the process begins with an intentional state – a conscious state of having as an object of thought some future state. Such an internal representation of the future is not a purely physical state at all, since the very idea of a 'representation' of something absent is not a physical property. To say that A represents B is not to state a property of A. A must represent B *to* someone, who takes A to represent B. The physical state of A is describable in terms of physical science – mass, position, momentum and so on. That it represents a desired future is really a state of the mind that thinks of it as such, and is not describable within the terms of physics.

Similarly, the idea of a future state as a value, as something worth choosing, ascribes a non-physical property to a state. In reflecting on how to realise the desired state, we think validly or invalidly, truthfully or untruthfully, appropriately or inappropriately. Considerations that are normative enter into this process essentially, whereas physical processes cannot as such be assessed as correct or incorrect, reasonable or irrational. Finally, in the realisation of the desired state there is the enjoyment of value, which is the final term of the whole process. But enjoyment is again a mental, not a physical, property.

In short, in personal explanations, there are elements of representation, normative assessment, evaluation and enjoyment, which are not parts of any physical process. Moreover, the process itself is non-algorithmic in that it follows no quantifiable, predictable universal law. It is not the case that from some evaluation and intention a predictable outcome always follows. The best we can come up with is a general statement that people are liable to aim at goals such as their own well-being or happiness. But all attempts in economics or political theory to predict exactly what those goals will be, or how they will be sought, have failed dismally.

So it seems that personal explanations refer to features of the world that are not physical properties. They are not capable of being put into the precisely quantified and deterministic forms that are characteristic of physical explanations. Physical and personal explanations are forms of explanation that are quite distinct.

9. CONSILIENCE

The only recourse for a naturalist like Wilson is somehow to translate the mental into the physical, to map mental states on to physical states of the brain, and then to explain events in terms of changes in those physical states alone. But that is impossible. To map mental on to physical states, you would need translation rules by which a mental state could be mapped on to a physical state, or at least on to a range of physical states. The problem is that very many thoughts are new and original, and so there cannot be an existent translation for them. Also they are rarely repeated exactly, since even prima facie qualitatively identical thoughts differ in specific mental content because of their differing locations in psychological history. So you can never set up the required translation rules. The mental cannot be mapped on to the physical.

It seems, then, that personal explanations are irreducible to physical explanations and have a substantial explanatory role of their own to play. They point to aspects of reality that are not physical, in the ordinary sense of possessing mass and volume, or momentum and position in space. Indeed, I have suggested that personal explanation is in one sense the more ultimate of the two, since a truly ultimate explanation must account for the existence of the otherwise unexplained ultimate laws of nature. The most satisfying form of explanation would be in terms of value, showing that it is a good thing, or that it leads to many otherwise unobtainable good things, that the laws of nature are as they are.

Science itself is committed to pursuing values such as those of truth, elegance, simplicity and constancy, in its quest to understand the universe. Mathematical physics has been well described as the poetry of the universe, and the value of understanding is the driving motivation of science. It is paradoxical that some distinguished proponents of such a value-driven discipline should aim to exorcise all values from explanation of the universe.

There are other values, of creative freedom, empathetic appreciation, and sensitivity, that also have a place in reality, and can only be grasped by the more participative forms of personal knowing that are largely found in the humanities. To understand a piece of music is not to show how it follows from previous states according to general laws. It is to understand the skills involved in imaginatively writing it, the pattern and form it has, the personal feeling for reality it expresses. To understand a successful play is to show how it captures aspects of human life in a distinctive way, how the words are chosen to evoke profound feelings, and how the plot achieves a satisfying formal shape.

Such forms of explanation are not reducible to physical explanations. It is perhaps harder to show that they reflect objective features of reality and that they point to a personal dimension to the world with which the natural sciences are not concerned. Yet if the cosmos shows intelligibility and elegance, and has a propensity to generate emergent beings capable of creative freedom and intelligent understanding, the existence of such a personal dimension might seem to be supported by the scientific enterprise.

Scientific understanding can perhaps show that there is beauty and elegance in the natural order, that there is a propensity in cosmic emergence towards life and consciousness, and that those elements of nature that seem harsh or indifferent towards human life are in fact necessary to the highly integrated structure of physical laws that make emergent human life possible. The scientific enterprise is not so indifferent to value as it may sometimes seem to be. The gulf that still sometimes exists between the sciences and the humanities can be bridged by a deeper understanding of the poetry of science, and of the truth-disclosing character of music, literature and art.

There can and should be consilience between physical and personal explanations. But it will not be a reduction of one to the other. Science is, after all, an activity of intelligent and creative personal agents. The physical explanations that constitute the study of nature may aim to be value free in their expression, but their pursuit embodies the values of intelligibility and understanding, of intellectual beauty and elegance, that arise from the intrinsic nature of the human intellect.

On the other hand, the goals that humans articulate and express in the exercise of their free agency require a regular, ordered and largely (but not wholly) predictable world as their context. So true consilience will show how physical and personal forms of explanation are oriented to each other and how each requires the other for a full understanding even of its own proper character.

10. ULTIMATE EXPLANATION AND VALUE

I have suggested that an ultimate explanation of our universe will have to include both scientific and personal forms of explanation, working in harmony to form one coherent worldview. The two main competitors for such an ultimate explanation are the reductive materialist hypothesis of E.O. Wilson's consilience, and the hypothesis of God, as a self-existent ultimate mind, creatively actualising this universe as a law-governed and intelligible unity for the sake of the distinctive values it makes possible.

A reason for thinking that Wilson's consilience hypothesis is unsatisfactory as an explanation is that it simply eliminates all the data of mental experience, purpose and value by translating them into purely physical terms. No one has ever completed such a translation, and I do not think it can ever be done, in principle. That leaves God. But for the God hypothesis to work, this universe would have to be plausibly seen as one that is necessary to the existence of a distinctive set of overwhelmingly great values. For many people, that is a major problem. There are too many disvalues – of suffering and personal frustration and sadness, for example – and the processes of cosmic history seem too accidental and fragile to be aimed at actualising a great set of values. To make the God hypothesis plausible, these issues need to be addressed. What I shall next seek to do is to show how the existence of human persons requires that the universe have the general character it has, with all its conflicts and ambiguities. Yet the existence of such persons is a great and otherwise unobtainable good.

11

THE EMERGENCE
OF THE SOUL

Persons are integral parts of the physical universe, and they arise by a continuous development of complex integrated systems in that universe. Yet they also have an immaterial or spiritual aspect, not reducible to purely physical properties. They are subjects of experience and action, 'souls', who might in principle be abstracted from their physical embodiment. In this respect, they share the nature of ultimate mind, though in a limited and physically dependent way.

AIM: To show that the idea of the soul is not of something quite distinct from the physical universe, and yet also not of something simply identical with part of it. The soul exists in continuity with nature, but emerges as a new level of reality that has unique value and dignity. The soul emerges from matter, but in its conscious life it shares the nature of the divine.

1. HUMAN PERSONS

Human persons have complex and varied mental lives. They are finite minds. If a theist gives mind ontological and causal priority over matter, then it is unlikely that finite minds should be wholly dependent upon and caused by the material universe. If mind – the mind of God – is the primary reality, it would seem that finite minds will not be wholly reducible to matter. Yet it also seems unsatisfactory that minds should be just injected into the physical universe at certain points, like alien beings in an otherwise continuous and gradually developing process. How can it be that finite minds seem to emerge from physical evolution and yet seem to be different in kind from matter?

I think a clue to understanding this is provided by the fact that consciousness requires objects to apprehend, and, as some interpretations of quantum physics suggest, objects are precisely objects-for-consciousness, not realities that exist entirely in their own right. So neither consciousness nor physical objects can exist in isolation.

Thus far in this book three main types of objects-for-consciousness have been mentioned, though that is not to say there could not be others, many of them unimaginable by us. First, there are the possible states of being, mathematical and evaluative truths, that are conceived by God. These are mere possibilities, not fully actual, but they must be thought of as part of the content of the divine mind.

Second, there are the selected and actualised contents of the divine consciousness, the contemplated contents of the divine mind itself. These are actualised from the general realm of possibilities as a set of states of very great value, of greater value than any set that could be possessed by any other being. They constitute the goodness and beauty of the divine being and establish that being as the Supreme Good.

Third, there are objects in the physical universe, the sensorium of God. This universe has an actuality distinct from that of the divine nature, a certain 'otherness' of being, with its own laws of development and its own principles of self-organisation and self-shaping. Yet the cosmos is still an object for divine consciousness, and without that consciousness it would not exist as an actualised object. In saying this I am committing myself to some form of Idealism and hypothesising that objects cannot exist without consciousness.

This hypothesis is in complete opposition to the materialism that is often thought to be an implication of modern science and philosophy. But in part 1 I argued that quantum physics undermines materialism and queries the very concept of 'matter' as a basic explanatory tool. Further, at least some weighty interpretations of modern physics seem to show that consciousness is necessary to the existence of the world as it appears to us. It is false to think that modern science is as such materialistic. Paul Davies and John Gribbin have even written a book on modern science called *The Matter Myth* (Viking, Harmondsworth, 1991). Consciousness may be an ultimate constituent of reality, and different sorts of objects may provide different sorts of contents of consciousness.

In addition to the three sorts of objects just described, the existence of finite minds adds a fourth sort of objectivity, involving not only a distinctness from the divine being but also a form of actuality which is qualitatively like the divine in one important respect. Finite minds are not only objects. They are also subjects with their own unique perspectives, their own creativity and their own ways of relating to other subjects in co-operation or in opposition.

2. THE EVOLUTION OF SUBJECTIVITY

We might see finite persons, including human persons, as both objects of divine consciousness and subjects of distinctive perspectives on the world. They have their own unique point of view from which they see things, and their personal experience is different from that of any other being. They have their own distinctive modes of creative action in the world. They respond and react to things in ways that express their own distinctive characters and experiences. They have their own unique ways of relating to other subjects within the world, so that they form a community of persons related to each other in many complex and changing ways.

These subjects are essentially embedded in the world, which provides the objects of their knowledge and response, and the environment of their actions. They are not independently existing pure spirits, but parts of an objective physical world within which they have their own subjectivity. So the objective world must be a suitable home for such unique subjects. In its gradual development over billions of years of complex organic structures with a huge degree of integration, of bodies with central nervous systems and brains, the physical world comes to be an appropriate home for many kinds of finite subjectivity.

That suggests a very important reason for the existence of a universe like this. The divine mind does not develop from anything else. It is the basis of everything else – of everything that is possible and actual. It does not grow in perfection, for it is perfect, omniscient and omnipotent from the first (this does not entail that God does not change at all, as will be made clear in chapter 14). It is immaterial and does not depend on the changes, instabilities and decays of any physical universe. It is, as Aristotle put it, supremely happy in the contemplation of its own beauty and perfection. But if it is good for the ultimate mind to be conscious of many forms of beauty, then to create other conscious subjects who can appreciate and create new forms of beauty is also good, and a new kind of good.

This universe is one that generates conscious subjects by a long process of development. These subjects develop from simple elements of matter, and they remain wedded to the material elements that form their basic structure. The sorts of experiences they have, the sorts of things they can do and the sorts of relationship they can have with others all depend on the structure of the physical universe of which they are part. Like God, they are conscious subjects. Unlike God, they are many, dependent and developing. In and through them, the physical world of objects becomes transformed into a world of subjects, conscious, intelligent, free and responsible.

In any explanation of the universe, the evolution of conscious subjects from a purely physical cosmos must play a major role. I consider that some form of an Idealist view offers the best prospect for an adequate explanation. One such explanation would be that an ultimate subject of consciousness brings into existence, out of the infinite reservoir of possibilities inherent in its own nature, a space-time universe that develops more and more complex and organised forms. Eventually, the universe generates out of itself, in accordance with principles that have always been embryonically present, innumerable finite conscious subjects that can experience and shape the universe from within. Thus the one primordial self is reflected and expressed in many partial forms in the physical world. The ultimate goal of this process might be that the many conscious subjects should somehow be re-integrated with the divine originator of the process. All their diverse experiences and actions might be embraced by ultimate mind, and ultimate mind could then be said to have completed its own creative journey into the finite and the physical.

Those who are familiar with the work of the German philosopher Hegel will find here an echo of his extremely ambitious system. But this is not purely armchair speculation. It is an attempt to take seriously the findings of modern science and see what sort of ultimate explanation of the universe they might suggest. Any attempted synthesis of this sort will be contentious. Some might say it is premature, that we do not have enough knowledge to justify it as yet. I think we can say, however, that it is at least as well substantiated by the sciences as is materialism, its chief competitor. And if bold hypotheses were not attempted, science would never have progressed. So, making due allowance for a degree of objective uncertainty and provisionality, this sort of explanation seems to me to offer the consilience of personal and scientific explanation and the integration of many diverse data within one coherent framework which a truly ultimate explanation requires.

3. THE BRAIN AND MATERIALISM

One view that is excluded by the approach I have suggested is extreme dualism, which makes mind and matter completely different from, and independent of, each other, and sometimes laments their combination. Such a dualism is found in Plato, and hints of it can be found in Descartes – who, however, most of the time had a fairly orthodox Catholic and Aristotelian view of mind and matter as closely integrated. He said, in a passage lamentably overlooked by some readers, 'I am not just lodged in my body like a pilot in his ship, but I am intimately united with it, and so confused and intermingled with it that I and my body compose, as it were, a single whole' (Descartes, Meditation 6, in *Discourse on Method*, trans. A. Wollaston, Penguin, Harmondsworth, 1960, p. 161).

Another excluded view is reductive monism, whether material or spiritual, which either reduces mind to matter, so that consciousness does not exist, or absorbs matter into mind, so that the material world is seen as illusory. Materialist monism is rarely found among the classical philosophers, who tended to take it as obvious that the data of consciousness are real and not extended in public space. But it has formidable defenders at the present time, and some neuroscientists take materialism for granted. This is because knowledge of the brain has increased enormously in the last century. I think it is fair to say that the neurosciences are still in their infancy. There is a huge amount yet to be learned, and there are many problems yet to be solved. But knowledge of the working of the brain has contributed hugely to our understanding of the human person and of the nature of mental activity. We can scan the brain to locate which areas of it are active when different mental activities are being undertaken. We can observe differences in behaviour and perception when parts of the brain are inactive or even absent. All evidence suggests that the sort of consciousness a person has depends on the functioning of the brain. Benjamin Libet claims to have shown by experimental investigation that even the initiation of voluntary action shows up as a 'readiness potential' in brain activity before an agent is conscious of making a voluntary choice (Benjamin Libet, Anthony Freeman and Keith Sutherland, *The Volitional Brain*, Imprint Academic, Thorverton, 1999). This suggests to some that it is not really the agent who initiates an action. It is the brain that really causes action, and the consciousness of 'acting' is an effect of prior brain activity. These findings are still controversial, and Libet himself believes that voluntary actions can still be modified or even blocked by a free decision of the agent (that is, he accepts the existence of radical freedom). But the accumulation of data like this from the neurosciences leads many to believe that the physical brain is the real cause of all actions, and that consciousness is just a by-product of brain activity.

At its most extreme, consciousness is said, by reductive materialists like Francis Crick (*The Astonishing Hypothesis*, Simon & Schuster, New York, 1994), to be *nothing but* physical brain changes. Other writers, usually more nuanced and sophisticated, occasionally fall into uttering such thoughts. Matt Ridley says that 'memory may consist, quite literally, of the tightening of the connections between neurons' (*Genome*, p. 227). On the next page, he admits that 'this is nonsense'. But what he then says is not much better: 'Memories ... reside in ... the neo-cortex.' This is rather like saying that Mozart symphonies reside in the compact discs that play them. Without the discs, we would have no way of retaining the sound of the symphonies. But there is all the difference in the world between binary codes on a compact disc and the sound of a symphony. Similarly, without a neo-cortex, we would certainly have no memories. But to say that conscious memories are 'literally' in the physical structure of the brain is to confuse conscious experience and what is during human lifetimes its physically necessary condition.

The view that consciousness is just some set of physical states is so counter-intuitive that, though it has impressed many by its technical sophistication, it has convinced few. Most of us are more certain of our immediate experience of being conscious than we are of highly technical philosophical theories that claim to show that consciousness is an illusion. Who or what is supposed to be suffering from the illusion?

The brain sciences certainly make us much more aware of the importance and complexity of the physical brain in influencing and constraining conscious experience and action. But to most conscious people it still seems that consciousness is a real and distinct form of existence, and essentially consists in a form of subjectivity that cannot itself be perceived directly as an object. Each of us has sensations, thoughts and feelings of which we are directly aware but which are not accessible in the same way to anyone else. In knowing this, we are aware of our own subjectivity. At the same time, that subjectivity is not isolated and cut off from everything else, as though it were imprisoned in its own solitude. It is inwardly oriented to, and requires the existence of, objects in a material world. Those objects provide the content of experience without which the subject would be a mere abstraction, an observer with absolutely nothing to observe. We might say that all objectivity exists for a subject (whether that subject is God or a finite being), and all awareness is of objects (whether they are conceptual, mental or physical). In the case we know best, the human case, the content is provided by the physical world, but a unique and distinctive perspective on that world and a creative form of action within that world are made possible by a personal subject of experience and action.

4. THE SELF AS SUBJECT OF EXPERIENCES

Those who believe that all conscious states are simply the effects of physical brain states tend to deny that a human person is just one personal subject of experience and action. Different parts of the brain are responsible for different parts of our conscious experience, and so it may seem that there is no immaterial 'self' or subject that somehow controls, supervises and integrates them all. The philosopher Daniel Dennett writes, 'There is no Cartesian Theater; there are just Multiple Drafts composed by processes of content fixation playing various semi-independent roles' (*Consciousness Explained*, Penguin, Harmondsworth, 1991, p. 431). Different areas of the brain – the visual, auditory and tactile receptors, for example – are involved in all experience, and they are all cobbled together by a 'center of narrative gravity', a shifting, evanescent, 'fictional' (p. 429) narrator of our life experience. There are lots of different bits of conscious experience, but the human brain weaves them (more or less successfully) into the narrative of 'one experience'. There is no real continuing self or subject of experience, no 'hopelessly contradiction-riddled myth of the distinct separate soul'. 'What you are is the program that runs on your brain's computer', and if the narrative persists, then you persist, in the only sense that is possible for a fictional construct like you.

What philosophers like Dennett are doing is deconstructing the self into lots of different bits of experience, just as David Hume did. But then, unlike Hume, they identify these bits of experience with states of the brain – and, hey presto, consciousness has been explained. The idea we have that there is one continuous experience, 'viewed' by a diaphanous self, is replaced by the idea that there is one constructed narrative of a life, which may be more or less coherent and continuous.

But who reads this narrative? Dennett is right in thinking that there can be more or less coherent narratives of experience, more or less well-defined senses of being a continuing subject of experience. But in the end, as philosophers like John Searle (see for instance *Intentionality*, Cambridge University Press, Cambridge, 1983) have continued to insist, the narrative is understood by someone, the experiences are registered and interpreted by someone within a continuing sequence of experiences, and what makes them belong to one more or less unitary narrative is that they are *my* experiences, the experiences of this person to whom no one else is identical. I am not just the program. I am the person who understands the program, and this is a difference of enormous significance. Denial of a conscious self seems mostly to be based on a dogmatic denial that there is any non-material aspect to human persons. This results in the assertion that there is nowhere a 'self' could be; there is no physical part of the brain that can be identified with a 'self'.

That is hardly surprising. The self, if it is a subject of experience, is not an object that could be observed or located among physical objects, of which the brain is composed. If we are to think of a self or subject, it is essential to distinguish clearly the ideas of subject and object. Objects are the contents of conscious experience; subjects are active agents who are aware of objects and can act upon them. Some philosophers would recommend that we speak only of successions of experiences, that we do not need to speak of any subject that has experiences. However, we speak of such mental activities as remembering, thinking, concentrating and imagining. These are precisely activities, not a passive registering of experiences. The reason for speaking of a subject-self is that we need an agent of our mental acts – even observing something fairly passively is an activity. The self is the agent who actively binds past, present and future experiences together as experiences of one consciousness.

The topics of mind, brain, consciousness and the existence of a self are among the most hotly disputed topics in philosophy. No view has commanded the assent of all informed participants in such discussions. To deal with the topic adequately would require a whole volume – a task I am happy to undertake, but not here. But it is not true, as Dennett sometimes says, that 'materialism of one sort of another is now a received opinion approaching unanimity' (p. 106). All those who believe in God have good reason to deny materialism in general, and therefore to be sympathetic to non-materialist accounts of the self. But in this book the most I can hope to do is to set out some principal alternative positions in order to locate the view I am taking of consciousness on the conceptual map.

The brain sciences have provided much new information on the causes, structure and limits of mental life. They have, I think, succeeded in showing that a radical dualism of mind and brain is implausible, given the close dependence of personality and mental activity on the physical brain. They have not succeeded in showing that mental events are nothing but brain events, or that mental events are wholly caused by brain events. The way is open to posit a non-physical aspect to personal existence and to speak of a subject or self who is the active agent in all conscious experience and voluntary action.

5. SUBJECT-OBJECT DUALITY

There are many pathologies of the self, many ways in which it can be disabled or fragmented by brain malfunctions. But the properly functioning subject is, and knows itself to be, one self who grows in knowledge and understanding, accumulates many sorts of experience, builds them into a unique unity of consciousness and at least sometimes acts freely and responsibly on the basis of that unique experience.

Anyone who accepts the sort of ultimate explanation of the universe I have described will think that ultimate mind, the source of all physical reality, is such a non-physical subject. But it will be unique in that it does not grow in understanding and it does not depend on any physical environment or brain. The primordial objects of its awareness, its ideas of all possible worlds and states, and the actual properties of its own being, will be parts of itself. It generates further physical objects and finite persons by its power of actualising possible states. So it is the one and only self-dependent being, upon which all physical beings depend for their existence. It is the ultimate subject.

Finite personal subjects, on the other hand, are wholly dependent upon ultimate mind. At least within this cosmos, they come to exist as integral parts of the total physical reality that is our universe. They can only come to exist in a physical environment that is prepared for them and that provides the possibility for a specific point of view and locus of action – that is to say, in an organic body in a social context with a well-developed central nervous system and array of sense organs. When such an environment comes to exist, a conscious subject naturally arises – somewhat as the basic forces of nature come into being in a specific situation micro-seconds after the big bang.

Since subjects are different in being from objects, the situation in which physical environment and conscious subject both exist could be called a form of duality, but it is a duality in which each part is necessary to the other. The misleading aspect of Cartesian dualism is that, in thinking of two seemingly distinct substances, it gives rise to the belief that the two substances, mind and matter, can very well exist on their own, such that their unity becomes a puzzle. Indeed, Descartes believed that animals were unconscious machines without minds, and that minds could exist wholly without bodies (at least he imagined that they could, though I doubt he actually believed they did).

That is why it may be helpful to think, not of mind and matter, but of subjects and objects, where subjects are not themselves physical but require the physical world, or something very like it, to exist, and where objects exist as objects-for-consciousness (primarily the consciousness of God, the primordial subject), not as primary and independent entities with their own inherent and mind-independent properties.

What is required, and what much recent work in neuroscience has begun to explore (though it is sometimes hampered by the influence of reductive materialism), is an investigation into the way in which consciousness and materiality are tied together into a unity in which neither loses its distinctiveness and yet they are not two quite independent sorts of entity.

Consciousness is not an 'entity' alongside others. At least in the case of physically embodied subjects, it is what may be called the 'interiority' of the physical, what it feels like to be an entity that appears to other subjects as a material object.

6. THE PROCESS PHILOSOPHY OF WHITEHEAD

The Galilean (more properly, perhaps, the Copernican) revolution taught us not to think of humans as the centre of the universe, but to see human consciousness as part of a wider cosmic totality, as part of the natural world. Influenced by this perspective, we might hazard the suggestion that all entities have a subjective and an objective pole. The subjective pole is what it is like to be an entity. The objective pole is what an entity looks like to others. These are tied together, since no finite entity exists without appearing to others – at least to God, the ultimate subject. And no appearances can exist without something that appears, something with its own character and set of potentialities. Everything that has independent existence has a subjective pole, but of course it does not have consciousness. Consciousness arises as a fairly high-level actualisation of the inwardness of active, dynamic entities in systematic interaction with one another.

The fullest development of a view of this sort is in the work of A.N. Whitehead (*Process and Reality*, ed. David Ray Griffin and D.W. Sherburne, Free Press, New York, 1978 [1929]). This is a complex and difficult work, and Whitehead would have been the last person to wish it to be a dogmatic system, but one of its major features is its setting out the development of mind and consciousness as a continuous process from the primordial structure of the material universe itself – a philosophical view that Bertrand Russell also held for at least two days.

A main attraction of the view is that it sees all physical entities from the simplest to the most complex as a continuum. So consciousness is not a sudden quite discontinuous reality that emerges unexpectedly on to the cosmic scene rather late in the day. It is a development from much simpler forms of the subjective pole of all physical entities.

All entities, from the smallest subatomic particle to the complex systems of events that constitute brains, are sensitive to their environment. All respond to stimuli by active behaviour, by realising dispositions to act in specific ways when those dispositions are activated by the environment in which they exist. Thus electrons have an electric charge, a disposition to react in a specific way when in the vicinity of another charged particle. This disposition is a property of the electron and is activated only when there are other charged particles near enough to make a significant impact.

Whitehead's suggestion is not that an electron 'feels what it is like' to have such a disposition or for it to be activated. But he does suppose that the electron has an inner state, quite unconscious, which is the state of having a set of dispositions, ready to be activated in response to the properties of other particles. It must not be forgotten that all such 'discrete entities' are 'slices out of overlapping and dynamic fields of potentiality' (p. 104 above).

As entities become more complex and form systems, their inner states change in character. At the level of biological organisms – say, in an animal body – cells in the body switch on specific developmental programs according to how they relate to other cells in the organism. All cells are identical in origin, but they differentiate according to their location in the developing body. The inner state of a cell is more complex than that of an electron. It has more dispositional properties, and it begins to make sense to speak of a cell as seeking or avoiding 'good' or 'bad' situations. Some cells will defend the body from attacks by a virus, avoiding a dangerous situation and seeking to re-establish the stability of the system. Whitehead calls such complex inner states 'feelings', but again this is not meant to connote consciousness.

However, consciousness does originate with the development of central nervous systems. Such consciousness is not a new and alien element added to the body. It is a development of the complex inner state of an integrated organic system. All such states develop in accordance with principles inherent in them. Most of those principles only become apparent when complex systems of events come into being. All have the capacity co-operatively to form successively higher patterns of complex unity that actualise new powers and new grades of inwardness.

So Whitehead would say that four main characteristics – of sensitivity to environment, reaction or inner adjustment to sensed stimuli, a dynamic response that realises some dispositional property of the inner state, and a propensity to interact within systems to form ever more complex unities – constitute the entities of which the cosmos is composed.

Without having to accept the whole Whiteheadian 'process philosophy', as it is called, I think it is helpful to adopt this continuum view of the development of consciousness. It helps to avoid both a radical mind–body dualism and reductive monism, and gives consciousness an intelligible place in the structure and development of the cosmos.

7. THE CONTINUUM OF NATURE

The Newtonian revolution tells us that the universe operates in universal law-like ways. So, however finite consciousness originates, it does so in accordance with universal principles that relate it in an intelligible way to the structure of the cosmos. Rather than thinking of mind as a novel and entirely different sort of being that is introduced into matter at a specific stage, we might think of consciousness as a higher-level actualisation of the primal sensitivity, active response and interactive self-organisation that characterise all the fundamental entities of which the cosmos is composed, in addition to their 'material' properties of mass, position and momentum.

The Darwinian revolution taught us that complex entities develop from simpler reactions and responses to a wider environment that 'selects' some forms for further development. In this way we could see consciousness as part of a continuum that progresses from the simple momentary unconscious sensitivities and responses of the fundamental events that lie at the basis of the material world, to the extremely complex inner states of organised spatio-temporally extended systems of events that characterise human organisms and brains – and perhaps to forms of life we cannot yet imagine.

As well as the world of solid coloured three-dimensional objects that constitute the world as it appears to us, there also exists a world comprising the inner states of complex and dynamic systems of events. Some of these are conscious states, and some develop that sense of temporal continuity and self-awareness that leads us to speak of a 'self'. Further, the fact of entanglement or interconnectedness of which quantum theory makes us aware suggests that such systems are not built out of simple independent atomic events. Rather, a system is an interconnected set of events such that the whole system constrains, without totally determining, the behaviour of its constituent events.

Building on these perspectives, we might think that subjectivity, or at least an interior structure that carries the potential for subjectivity, is built into the structure of the physical from the first, but develops towards conscious response to stimuli and then towards self-awareness and intentional action as the interconnected, dynamic and developing physical structures of the universe grow more complex.

The picture is that finite consciousness is not a sudden unexpected intrusion into a universe that has been purely material up until that moment. It is a development of properties that have been potential in the material universe from the first, but only gradually unfold as material systems grow more complex and organised.

Such a scheme builds conscious subjectivity into the universe in a more integral way than Cartesian dualism seems to do. This scheme is not a form of materialism, for it affirms the existence of phenomenal properties, of intentional causality and of emerging purpose and value, as real constituents of the universe. Nor is it a form of reductive Idealism, for it affirms the independent reality of the physical, which provides the content for embodied minds, and whose emergent development lays down the causal parameters of knowledge and action. The existence of conscious selves can be seen as a fairly high-level stage in the development of the inner states of complex systems of events, systems in responsive and dynamic interaction with the total environment of the cosmos.

8. THE EVOLUTIONARY STORY

On this view, even the simplest object has some minimal inwardness and potency, though it is not conscious and it will be just about completely constrained by the laws and powers of its environment.

At a higher level there will be some sensitivity to external influence, and some form of primitive response, perhaps of growth or reproduction. Plants feel the sun; they move towards the sun. But this is a matter of physical stimulus and response. It is misleading to think at this stage of inner feeling or intentional movement.

As physical structures get more complex, and central nervous systems and primitive brains form, rudimentary consciousness of the environment arises. Environmental stimuli are apprehended as qualia (phenomenally cognised properties or states). The simplest conscious subjects probably just register colours, shapes and smells in a particular space, as they are provided by their sensory receptors and processing unit (the brain). They may feel pleasure or pain, and act to seek pleasure and avoid pain, in limited ways. They have no sense of continuance, but are momentary.

Momentary feeling-states and reactions are new sorts of reality, elementary and momentary subjective states, 'sparks of consciousness'. They build upon and develop the existing physical sensitive-reactive system in new ways. In that sense they can be called emergent phenomena.

Such elementary subjective states can be connected up in a series where past states enter into the interpretation of present states and look towards a future. At first this may still be a series of short chains of thoughts, feelings, sensations and percepts. There need be no sense of responsible action, no abstract reflection, no imagination free from sensory presence, no calculation of the thoughts and feelings of others (and so no sense of personal relationship). This is a primitive sense of continuing self and responsibility, without clear awareness of it, or a strong sense of personal identity – perhaps most higher mammals possess this form of consciousness.

9. THE CONTINUING SELF (THE SOUL)

A further stage in development is characteristic of humans. It involves a degree of responsible self-shaping and free conceptual thought. Language seems to be essential at this stage. There arises a sense of moral community and a sense of continuing self.

At this stage there is a reflective use of the past to interpret present experiences, to classify and recognise them. Present experiences are evaluated as desirable or undesirable, to be sought or avoided. There arises the anticipation of future experiences, and propensities to respond are given direction in accordance with what is desired for the future. The physical lays the basis for these feelings and acts. But now organisms may feel the physical stimulus, registering its quality and intensity. They may act by modifying their natural physical responses, formulating a goal that may orient them more exactly. So a continuity of experience and agency arises for which there is an idea of one continuous experience, the past interpreting the present and giving rise to the goal-directed realisation of future states. This continuity is not necessarily temporally continuous. There can be breaks, as long as the past is used to interpret the present and the present is used to anticipate the future, so that there is a chain of experiences, connected by memory and anticipation.

There could be transient unities existing only for a moment, then vanishing. There could be short-lived, intermittent or divided unities – short snatches of memory, mixed with error, so that the subject-self that unites experiences into one would be forgetful and inconsistent in goal-directed action. These would be imperfect unities, with little or no sense of continuity, no sense of continuing self. But the ideal unity of consciousness would be a comprehensive, complete and wholly understood set of experiences, connected by perfect memory and anticipation, perfectly controlled in action, always seeking coherent and clearly conceived goals. This may occasionally exist among humans, but not very often. It is a state largely yet to be attained, but a possible one for human beings.

The unitary identity of consciousness is thus a matter of degree. We have experiences, but do not fully understand them. We remember, but forget much. We anticipate obscurely and act indecisively and with incompatible goals. Still, we have a sense of a continuing and more or less responsible self, with experiences we can remember and past actions that have partly shaped us and our dispositions. This is the stage at which it is possible to speak of a distinct kind of subject of experience and action, not necessarily confined to members of the human species, but marking off a distinctive class of persons. Such a subject of experience and action is what has, in traditions influenced by Aristotle, been called an 'intellectual soul'. (A discussion of this topic appears in Keith Ward's *Religion and Human Nature*, Clarendon Press, Oxford, 1998, chapter 7.)

According to this evolutionary story, the intellectual soul is in a continuous line of development from the simplest physical particles. It is the inwardness of complex integrated systems of events. But, though there is continuity in the development of such systems, there are also crucial phase changes in the process. An example of a phase change is the change from liquid water to steam, as the temperature gradually rises, and the change to ice as temperature falls. The rise or fall of temperature is continuous and gradual, but at crucial points there is a sudden transformation of matter into a different state.

10. KEY PHASE TRANSITIONS IN EVOLUTION

In the process of cosmic evolution, there are four particularly important phase changes in the structure of matter. First there is the point at which the dynamic and momentary fluctuations of fundamental particle/waves stabilise into more enduring atomic structures, and the chemical elements of the periodic table establish new patterns of interaction between these atoms. Second is the formation of systems of macro-molecules like RNA and DNA that can replicate and instruct proteins to build carbon-based bodies. Third is the genesis of consciousness, as central nervous systems emerge within organic life forms. And fourth is the emergence of the sense of a continuous, intelligent and intentional self in humans (and possibly in other life-forms, though we cannot be sure of this).

At each point a new and more complex set of properties comes into being, with new laws of interaction between them. The fourth transition is of particular importance, since it introduces a form of subjectivity that has, at least in theory, the ability to transcend the immediate context of its physical origin.

It makes no sense to speak of atoms existing apart from the physical elements (sub-atomic particles) of which they are composed. It does not make much sense to speak of DNA apart from the atoms that constitute it. However, even at this stage it is possible to speak of the 'code' that DNA carries for building organic bodies, and that informational code can be abstracted from its 'hardware', its physical carrier. When we describe the genome of a nematode worm, we can do so without having its physical bits in front of us. We can distinguish the information content of a system from its physical structure. It might be possible to embody the same code in a different physical structure – like moving a computer program to another computer. Nevertheless, we do not seriously think of the code existing apart from the nucleic acids in which it is embodied, or having some sort of enduring life of its own, except in a metaphorical sense. The distinction between information and physical structure is a purely conceptual one, and we would not normally think of sequences of information as having real existence on their own.

With consciousness, for the first time it does make sense to think of conscious states as having an existence that is in some sense independent of the publicly observable physical states in which they are normally embodied. As David Armstrong says in *A Materialist Theory of Mind* (Routledge, London, 1968), we can imagine conscious states existing without any associated physical states (though, as a materialist, he obviously thought they do not). But where conscious states are very closely tied to physical stimuli and where there is little or no sense of continuity, of being 'the same consciousness' throughout a stretch of time, there seems little point in thinking about such conscious states being abstracted from their immediate physical context. Any such consciousness would not be aware that it was the same consciousness that used to exist at some other time or place. There would be little conscious memory or hope and fear for the future. Without a sense of temporality and continuity, a sense of self would not arise.

So it is only with the fourth transition that it fully makes sense to speak about the existence of a free and responsible self, which is what has in the Western tradition been called an intellectual soul. And indeed we hold mature and normally functioning human adults, but not dogs and cats, responsible for their actions. Human persons react to the events they experience in complex ways and gradually build up a personality by their responses and activities. They reflect upon their experiences and construct imaginative worlds removed from the actual world they inhabit. They form moral communities in which principles for action are decided upon and reinforced and within which a sense of moral obligation and of social responsibility develop to various degrees.

Humans have an imaginative, creative and partly self-shaped inner life. Although rooted in a particular body and social and physical context, their mental life can to a limited but real extent roam free of attachment to that context. When the ability to act on general principles and to accept moral responsibility for one's actions arises in physical organisms, we can speak of an enduring subjectivity that is self-consciously the same throughout all an enduring subject's experiences.

The sense of being a continuing reflective and responsible self, living in a moral community of similar selves, within which discussion and the formulation of general principles of action take place, is one that, so far as we can ascertain, exists on this planet only among members of the human species. That is the basis for saying that humans are of unique and distinctive moral value and that they have a special claim upon our respect. They are nevertheless integral parts of this complex and developing physical universe, and, since their development is an important part of what they are, they could not have come to exist in any other world than this. It seems that they could logically continue to exist in other forms of embodiment. Whether or not they do so is a question to which I will return in chapter 17.

12

THE ORIGINS OF CULTURE

The human person exists in a society in which art, morality and religion, as well as the understanding of nature given by science, emerge as aspects of personal life which can exhibit intrinsic value and which require and express distinctive personal forms of knowledge. In particular, the emergence of moral responsibility introduces quite new dimensions of reality and value into the cosmos.

AIM: To show that morality grows from the development of genetically selected dispositions towards aggression and altruism, but allows a real, if limited, choice between them. To show that an evolutionary account gives a good explanation of why humans feel moral obligation and yet often behave selfishly, of why they feel morally free yet often enslaved by their desires.

1. PERSONS AND SOCIAL CULTURE

Persons are not isolated private subjects of experience, cut off from any real contact with others and with an outside world. This is another respect in which the philosophy of Descartes has been used (wrongly, in my view) to support a radically individualist, even solipsistic, ideology. Personhood does carry important elements of privacy and interiority, which must be respected. But since Hegel and Marx it has been more clearly recognised that personhood is importantly relational. We learn our language, our behaviour patterns and our fundamental values from the societies of which we are part. Our way of looking at the world and our basic attitudes towards it are not wholly private and incommunicable. They are formed in conversation and interaction with others and are parts of a continuing history and social tradition. Even the most private aspects of personal life take publicly accessible form in the artefacts of culture, of literature, art and music. Such works give a real, if indirect, knowledge of the subjectivity of their creators. I do not mean that we gain a magical entry into some purely introspective experiences. I mean that we can learn to apprehend and feel the world as writers, artists and musicians do, to experience in ourselves the unique perspectives on the world that their skills enable them to express. Through training and experience, we can come to feel what it is like to experience the world as they do, though we always do so from our own perspective, which adds another layer of meaning, making this form of knowledge intensely personal and only indirectly communicable.

Learning to appreciate the arts involves training in discernment, empathy and imagination. It requires participation in a social life in which knowledge is imparted and shared and creative projects are fashioned in co-operation with others. In this way social culture enables a development of what is most truly personal, of forms of knowing that are otherwise unobtainable, and of forms of social life that are of intrinsic worth. Within such a culture a moral sense begins to form, as empathy and participation bind us into a wider world of persons, which gives our lives meaning and significance.

Cultural artefacts can be seen as manifestations of distinctively human excellences, capacities of discernment, evaluation and imagination that set humans apart as beings of value in themselves. Both the production and appreciation of cultural objects require training in skills of creativity and discernment, and a prioritising of the world of personal experience and responsible creativity. In such a world humans become, for the first time, moral agents.

2. SCIENCE AND PERSONAL KNOWLEDGE

It may seem that such a form of knowledge is quite different from the sort of objective, impartial and dispassionate knowledge that is sought in the natural sciences. But the values of creativity and discernment were given new emphasis and clarity with the rise of the natural sciences. The new sciences stressed the role of the creative scientific genius, like Newton 'voyaging on strange new seas of thought, alone'. They stressed possibilities of changing the world for the better, by creative effort. They stressed the importance of attending to the particular and the individual, instead of simply theorising about abstract essences. They stressed the necessity of cultivating a sensitive understanding of the world and of striving for truth despite the opposition of the crowd.

Far from leading to a view of the world as a value-free, coldly rational, mechanistic and purposeless concatenation of atoms, science was one of the major forces leading to Romanticism, to the celebration of lonely genius, of the awe-inspiring mystery of the world, and of the excitement of novel views and possibilities, new and ever-changing landscapes of the mind.

New discoveries about the wonders of the natural world and the ability of the human mind to discern them helped lead to an emphasis: (1) on creative individuality, the fullest use of all distinctively human capacities; (2) an emphasis on the importance of the emotionally charged insights of personal experience on the basis of which so many scientific discoveries were made; and (3) an emphasis on the formation of a society in which such capacities and insights could be shared as fully as possible.

A stress on the importance of creativity, of affectively charged personal experiences and of their development by free participation in a complex social culture, is the hallmark of that concern for human development and self-realisation that marked the eighteenth-century Enlightenment and its transition to nineteenth-century Romanticism. These are not anti-scientific, purely poetic, values. They are the values that lie at the heart of the scientific enterprise. They bring out the fact that scientific investigation, artistic creativity, moral commitment and religious imagination are not competing or conflicting aspects of human life. They are diverse expressions of the distinctive human capacity for uniquely personal forms of discernment and understanding.

Such values and capacities were not newly invented in nineteenth-century Europe, though they took new forms and definitions then. Their origins lie in early prehistory, when the mind first discovered its ability to rise to creative thought and reflective understanding, to discover in itself the reality and value of the personal.

3. COGNITIVE FREEDOM

There could not be culture without freedom from the constraints of directly adaptive behaviour, without the mental skills to remember, imagine, plan and create and without a desire to produce artefacts that will express and record for future societies something of the inwardness of personal experience, the self-conscious recollection of affective response and the striving for excellence that characterise personal existence. It seems that culture arises with the development of a sense of reflexive personal identity, and thus with a decisive phase change in the development of the human brain.

It is natural for a sense of the significance of the personal to arise with such a development. When persons are given significance, they begin to exert moral claims on others, and a sense of moral obligation emerges. It is natural to locate such obligations outside yourself, in objective reality. For most early societies of which we are aware, the sense of obligation is strongly linked to a wider sense of transcendence. The sense of the personal is extended, at least by some, to embrace the whole of the humanly experienced world. Religion arises as a natural but perhaps not universal corollary of cultural and moral activity. It has its own proper sphere in the attempt to achieve conscious relationship with a transcendent personal reality.

In early times, this is often conceived as interaction with one or more transcendent personal powers (spirits) in order to gain good and avert evil. Grounded in the riotous imaginative world of dreams and visions, these notions develop by various paths towards the idea of one supreme Spirit that embodies in itself the Supreme Good. In the classical religious traditions, this Spirit is believed to ground human moral obligations in objective reality, to aid humans to respond to them, and to promise the eventual realisation of a fully moral order, whether in this universe or beyond.

Within such a view of a transcendent personal or spiritual reality, the artefacts of culture become more than expressions of human imagination and creativity – though they are that. They are also expressions of spiritual culture, symbols of spiritual reality. They can, and many believe that they do, communicate something of that reality to those with eyes to discern what they express. In this way art, morality and religion express dimensions of personal being to which the natural sciences do not have direct access. They express a crucial transition from the world of the physical to the world of spiritual reality.

Attempts by evolutionary psychologists to explain artistic, moral and religious beliefs in terms of fitness for survival, as stable evolutionary strategies, often neglect the most important element that really binds such beliefs together. That is the evolutionary phase change by which the mental life of humans achieves a limited but real freedom from its biological bases in pre-conscious behavioural strategies. Only when we see cognition, imagination and feeling as taking on a distinctive life of their own, apart from the requirements of biological survival, can we discern what marks persons off from other forms of physical being, and what makes them closer in conscious relationship and in kind to the trans-physical source of all being, to ultimate mind.

4. FREEDOM AND DESIRE

Human persons are to some extent free of the forms of physical causality that gave them birth, and free to develop their futures as they decide. Such freedom is a great value, and essential to personal existence. But freedom is dangerous, and on this planet its misuse has led to the violence, warfare, exploitation and tyranny that is the tragic history of humanity.

Humans might have grown in sensitivity, creativity and companionship, gradually extending the disciplined control of compassion and empathy over the behavioural tendencies to aggression and ruthless competition that had brought them to dominance on the planet. But if this was a possibility, it was one they were not to realise.

Although the first millennia of human history are inaccessible to us, it is clear from some of the first archaeological evidence we have – evidence found in early graves of violent death, human sacrifice and cannibalism – that violence and cruelty had become common in many human societies in late Palaeolithic times. We can see how, over many generations, the pursuit of self-interest, or of narrowly tribal interest, could become entrenched in and supported by the social practices of human societies. This led to the systematic social sanctioning of aggression and exploitation. It came to seem morally right to protect your own interests and to bring all other peoples under the 'protective power' of your own authority.

Over long periods of time the basis of social morality has been insidiously corrupted, so that now no one is born outside the inescapable influence of greed and aggression. Almost everything we learn and imbibe from our societies encourages us in greed – from the 'legitimate' desire for a larger car to the provision of cheap clothing at the expense of exploiting subsistence workers in far-away parts of the world. Almost everything in our society depends upon the use or threat of aggressive force – from the institution of armies and police to keep the peace to the encouragement of competitive ambition to succeed in business.

These things are not wrong. Our societies could not exist without them. Yet they subtly corrupt the moral sense that knows it is right to seek the welfare of all humanity and wrong to caricature the beliefs of others or to make an exception of oneself. Such ideals come to seem Utopian and unrealistic. Though the world continues to be propelled towards its final destruction by the twin forces of greed and fear, there seems no way in which we can escape their grip.

There are echoes here of the Darwinian picture of evolution as a struggle for scarce resources, in which the weak are eliminated by ruthless competition – a picture that reflects and possibly derives from the laissez-faire capitalism of Darwin's own Victorian society, as well as from the gloomy forebodings of the Anglican clergyman and economist T.R. Malthus. To some it even seems that this is the natural way of things. Economic and class struggle to the death is how the human world operates, and we must either go with the system or be eliminated. It is not surprising that Darwin's theories were held by some to lead to social Darwinism, recommending the elimination of the unfit and unbridled competition to weed out the weak and produce a strong and healthy future for humanity.

I have already suggested that to see the evolutionary process as wasteful and cruel may be a case of the pathetic fallacy – the fallacy of attributing human feelings and thoughts to natural processes. The Darwinian account provides a good explanation of why greed and aggression are genetically selected tendencies in any successful species. They are not inexplicable temptations for perfectly moral and happy primitive beings. They are parts of the human genetic heritage.

But it is a mistake to think this shows that such tendencies are inescapable, or that they are the only or even the dominant tendencies likely to be selected in the course of evolution. It is an even worse mistake to think that morality is wholly determined by the selection of behaviour that proved adaptive in the distant past but has become catastrophically maladaptive in the modern world of nuclear and biological destructive capability.

5. THE BIRTH OF MORAL CHOICE

For many traditional religious views, humans originated in a state of what Hegel called 'dreaming innocence', not subject to overwhelming desires of lust and aggression, hatred and greed. They had sufficient awareness of good and evil, and sufficient natural ability to do what is right, that they might never have fallen into the grip of the egoistic destructiveness that is so obvious in the human world today.

We do not know enough about human origins to know whether this is so or not, and it is hard to think of any scientific evidence that could be found to decide the issue. There is plenty of biological evidence to show that lust and aggression would have been helpful characteristics in bringing a species to dominance on earth. So we might expect these to be natural, almost inevitable, traits of dominant intelligent organisms. On the other hand, equally helpful traits would be those of co-operation and self-sacrifice, at least within a small community of physically weak but intelligent individuals.

When Darwin wrote the *Origin of Species*, he had no idea of the part that genes play in heredity, or of the structure of DNA. Since the discovery of the double helix structure of DNA in 1953, and realisation of the possibility of identifying genes for elements of human behaviour or at least for tendencies to behave in certain ways, Darwinian theory has been applied to human nature in an attempt to explain why humans are the way they are.

The general theory that genes mutate and are preferentially selected by the environment has been applied to human culture, art, morality and religion. It has been postulated that there are genes for both aggressiveness and co-operation, for both sexual promiscuity and reserve. Some behavioural dispositions are more conducive to the survival of a species, and therefore to the propagation of the genes that cause them, than others. So if, for instance, there is a gene or set of genes for aggression, aggressive animals are likely to eliminate competitors and procreate more effectively, so the genes for aggression that they carry will be propagated more widely. The species will accordingly get more aggressive. We can then explain why we, the descendents of early members of such a species, are aggressive by pointing out that aggression was a good survival strategy millions of years ago, so we inherit genes for aggression now.

Many traits of modern humans can be explained as the results of successful evolutionary strategies early in human or pre-human evolution. The genes we inherit make us aggressive, and we have to do the best we can with that. We cannot ignore or eliminate aggression. It is part of our nature – not directly created by God, but developed by a combination of genetic variation and natural selection over millions of years.

6. ALTRUISM

For some years, those who held this theory were troubled by the fact that it did not seem to allow for altruism and self-sacrifice in human behaviour. And yet such dispositions exist. Can this be explained in evolutionary terms?

Illumination was to come from an unexpected source, the theory of games, a highly technical mathematical theory developed in its modern form in connection with models of economic activity (the seminal text is John von Neumann and Oskar Morgenstern's *The Theory of Games and Economic Behaviour*, Princeton University Press, Princeton, 1944). Game theory sets up a number of competitive situations – 'games' – and attempts to work out the best strategies for winning such games. It has become a sophisticated mathematical discipline that can often pick out the strategy that will enable one competitor to win most games in the long run.

Evolution can be seen as a sort of game, in which animals compete for scarce resources. R.C. Lewontin first applied game theory to evolution in 1961 ('Evolution and the Theory of Games', *Journal of Theoretical Biology*, pp. 1, 382–403). Others have since used it to show how co-operation can arise from the repeated application of competitive survival strategies (see especially Robert Axelrod's *The Evolution of Co-operation*,

Basic Books, New York, 1984). We might think that ruthless self-ishness is the best strategy for survival. But game theorists can show that this is not the case. For if two ruthless egoists meet each other, the likelihood is that they will eliminate each other, while more prudent by-standers will survive by default. It turns out that one of the best strategies, in general, is one sometimes called 'Tit-for-Tat'. Each competitor starts by being nice to the other, but if one defects and does something nasty, the other must respond in kind. No strategy is perfect, and this one can lead to terrible escalations of violence. Nevertheless, most players will agree at some point to start being nice to each other again, in order to escape an unending series of escalations. A sort of equilibrium will be achieved. Over many generations, and with many players, this is much more likely to happen.

Game theorists have devised many more complex strategies for 'winning' – for maximal survival – that can be worked out over many generations and in varying circumstances. Various strategies will be successful in various sorts of environment, and against various sorts of competitors, so it is difficult to lay down an infallible rule for winning. Nevertheless it is a general feature of the best strategies that a good deal of co-operation, though of a tenuous and limited kind, will be more conducive to survival than short-term egoism.

In this way game theory shows what strategies will lead to the preferential survival of any entities, even when such strategies are adopted unconsciously and by chance. In a highly competitive struggle for survival, a successful strategy will not be maximum selfishness on all occasions. Animals that co-operate in groups will succeed – will pass on more of their genes – more often than animals that practise rigorous self-interest.

Matt Ridley, in his fascinating book *The Origins of Virtue* (Viking, London, 1996), gives the example of worker ants that are infertile and give their lives for the sake of the queen and other ants in the nest. This is a successful strategy for the survival of the genes of their siblings (who inherit the genes of the same queen). So individual self-sacrifice is a successful strategy for the group. If you have a gene that switches on self-sacrifice for some individuals at some times, that gene (possessed by all members of the family) is likely to be passed on very successfully and widely propagated by the rest of the group. Limited self-sacrifice and co-operation is a successful survival strategy, and so it is likely to be genetically programmed into future generations. In the evolutionary game, the winners may not be ruthless egoists. They could be initially kind, sometimes self-sacrificing and co-operative within kinship groups, but they will also be aggressive and greedy, especially to outsiders with whom they do not need to relate over long periods of time.

The evolutionary account gives a plausible explanation of how it is that humans are selfish and aggressive, altruistic and co-operative, at the same time. Humans were not created perfectly virtuous. Nor were they wholly selfish aggressive brutes. From the first, they bore within them the inherited traits of both aggression and co-operation, and these evolutionary imprints of their successful past formed the basis of their behaviour. It was on this ambiguous basis that morality, the conscious rational direction of action on principle, developed.

It developed in societies within which empathy and co-operative creativity became fundamental to a new form of personal and social existence. The birth of morality is coterminous with the birth of the relational self. And that is an occurrence of such value and significance that the whole previous existence of the cosmos, though it has indeed its own beauty and value, can also be seen as preparing for the emergence of rational and responsible consciousness.

13

COSMIC ONTOGENESIS

Sociobiology gives a good explanation of the genesis of the human moral sense. It can be employed in a reductive way, implying that morality is an 'illusion', implanted by genetic imprinting. But the genesis of moral sense can also be seen as the emergence of a genuine knowledge of objective goodness. For the theist, this will ultimately be knowledge of one Supreme Good, and it will enable the transformation of human lives by that knowledge.

AIM: To show that evolution on earth can be seen as preparing embodied minds for knowledge of God, as the Supreme Good. To show that moral knowledge is knowledge of the Supreme Good, which exists as an actual and supremely desirable being.

1. MORAL AGENCY

The sort of conscious morality that human beings possess is more than a semi-automatic struggle between competing inherited tendencies. A fully responsible moral choice only exists when an agent is able to choose between two actions, knows what the actions are and knows the difference between right and wrong. These principles, known as the McNaughton rules, form the basis of ideas of criminal responsibility in English law. They, or principles like them, form the basis of much criminal law throughout the world. Human agents are morally free when they have developed the ability to evaluate possible actions consciously, and when no power outside their own wills determines what action will be taken – though many factors, including inherited dispositions, will influence their decision, sometimes very strongly. In those cases, however, we often speak of 'mitigating factors', to be taken into account in assessing responsibility.

We can imagine early humans facing a conscious choice between aggressive and co-operative action – say with the question 'Should I overlook my opponent's aggressive act or should I respond in kind?' An early form of game theory could be played, weighing up various possible consequences. But in the end the agent must just decide what to do. There are inherited dispositions tending in both directions, but let us suppose the agent comes to believe that one option is the morally right one – and suppose this means the agent thinks that this option will produce most good, regardless of any desire for revenge or personal welfare. Suppose in this case that forgiveness seems the morally right course, though it would be to the agent's advantage to kill the opponent. Both acts will be motivated by deep inherited tendencies, and may lead to pretty evenly balanced outcomes with regard to such things as the long-term survival of the tribe. So both acts will be more or less equally rational. Yet now the agent can choose between right and wrong, and nothing makes one of the choices inevitable.

This basic moral choice is between goodness for its own sake and personal advantage driven by desire. The course of evolution on earth has led to making such a choice possible and intelligible. It helps to provide an intelligible background for understanding human moral choice. But humans cross a decisive threshold when they are able to make such a choice. This is what makes humans distinctive among most, and perhaps all, other animals – they can choose to act out of a conception of the good, for love of the good alone.

It is because so many humans have so often chosen in favour of selfish desire rather than goodness that passion, greed and violence now seem to rule the human species, and goodness seems so difficult. The present human condition is not merely the inevitable result of a ruthless and competitive process of evolution that has rendered humans helpless in the grip of primordial biological forces. It is the result of a past history of perhaps four or five million years of free and conscious choices of personal passion and advantage, reinforced by generations of cultural training so that destructive and egoistic choices have now come to seem natural or even inescapable for human beings. Humans, with their capacity freely to choose beauty, understanding, goodness and love, have instead released the negative possibilities of being – ugliness, ignorance, destructiveness and hatred. Humans still have these choices, but their freedom is impaired by the negative cultural inheritance that now makes genuine goodness rare and difficult and often seems to enslave them to selfish desire.

Humans are, however, never wholly enslaved. As Richard Dawkins memorably put it, 'We alone, on earth, can rebel against the tyranny of the selfish replicators' (*The Selfish Gene*, Granada, London, 1978, p. 205). Humans have the capacity to escape from their inherited predispositions and to choose the good for its own sake. For the theist, that is not just an accident of evolution. It is human destiny.

2. GENETIC SELFISHNESS

Some evolutionary biologists seem to tell a different story. Anthropologist William Irons writes that 'the central tenet of socio-biology is ... that human beings are bundles of inclusive, fitness-maximising mechanisms, shaped by a history of natural selection' ('Morality, Religion and Human Evolution', in *Religion and Science*, ed. W. Mark Richardson and Wesley J. Wildman,

Routledge, New York, 1996, p. 394). Humans are mechanisms, and their main function is to propagate genes. Morality does not figure very largely in this view of human nature.

'Our brain and sensory system evolved as a biological apparatus to preserve and multiply human genes', writes Edward Wilson (*Consilience*, p. 55). Even our noblest aspirations and our deepest thoughts are just instruments to multiply genes. All I can say is that, in that case, most of mine have been a complete waste of time. We might wonder whether a serious concern for truth will survive such an estimate of human intelligence.

Matt Ridley goes even further and says, 'Altruism is just genetic selfishness' (*The Origins of Virtue*, p. 19). Now it looks as if our highest and most self-sacrificial moral acts are mechanisms for multiplying genes. Morality itself is a sort of illusion. The danger is that this will be misunderstood as saying that we have to be selfish, that true self-sacrifice for the sake of the good is impossible and that it is senseless to talk of human responsibility or morality. That, I think, is not what the writers have in mind, but their language is misleading.

For them, what Darwin's theory shows is that 'the human being [is] just another animal ... the disposable plaything and tool of a committee of self-interested genes' (Ridley, *The Origins of Virtue*, p. 19). Impressed by the close genetic similarity between the great apes and humans, they overlook the vast difference between animals that can consciously envisage, discuss and jointly decide upon general principles for action, and animals that simply act on instinct, or genetically programmed routines and direct response to perceived sensory stimuli. Of course there will be intermediate cases in evolutionary development, and the great apes almost certainly are such cases. But there is a difference in kind between a being that can act in order to realise a consciously envisaged goal, that can reflect upon what course of action it ought to adopt, and a being that plays out an implanted programme of action.

Some evolutionary thinkers apparently deny this difference in kind. E.O. Wilson, the founder of sociobiology, now usually called 'evolutionary psychology', claims that 'Ought is just shorthand for one kind of factual statement, a word that denotes what society first chose (or was coerced) to do, and then codified' (*Consilience*, p. 280). Some principles of social action and organisation are genetically programmed (as are the roles of worker ants). They have been conducive to the survival of certain societies and so they continue to exist in those societies. They, Wilson suggests in this extreme mood, just *are* our moral principles. He confirms this extreme view when he says that 'rational choice is the casting about among alternative mental scenarios to hit upon the ones which, in a given context, satisfy the strongest epigenetic rules' (p. 199). An epigenetic rule is a genetically coded rule, built up over millennia of selection for survival. It may seem that we, or our ancestors, rationally chose some principles of action. But the fact is simply that the strongest genetically coded rules came to dominate our lives and now appear to us to be principles that we 'ought' to follow. Reason has no real part to play in the development of moral principles, since our moral principles are just what evolution has programmed into our genes, the real determinants of human acts. Any appeal to conscious rational choice is avoided, and individuals 'do things that benefit their genes' (Ridley, *The Origins of Virtue*, p. 17).

This replacement of free moral decision by genetically determined behaviour does not help us to make moral decisions. In fact it suggests that we are never free to make such decisions, and that we will do what we are genetically determined to do anyway. But the obstinate fact is that we do have to make important moral decisions, and so we must act as though we could choose between alternative possibilities, as though we were not determined. No allegedly scientific statement about what happened in the past can make decisions for us, and no description of successful evolutionary strategies in the past can take out of our hands the responsibility for deciding what to do now.

The mistake that William Irons makes is to identify the evolutionary history that has shaped human beings to be the way they are with what human beings actually are now. It is to say that human beings are nothing but efficient replication mechanisms. But though they are such mechanisms, just as they are bags of water and chemicals, they are much more. When intelligent agency arises in such 'machines', there also arises the possibility of a conscious selection of future states because such states are considered to be of intrinsic value. At that stage, it may well be that reproduction is no longer considered to be the only value. It may not even be considered a decisively important value, for humans may, and sometimes do, choose death rather than dishonour, disloyalty or injustice. When that happens, an account of morality solely in terms of reproductive efficacy, whether past or future, is grossly insufficient. Edward Wilson helpfully reminds us that we should not place our moral demands beyond the realms of possibility. He says, 'It should be possible to adapt the ancient moral sentiments more wisely to the swiftly changing conditions of modern life' (p. 285). We must not ask of people what is impossible, and it is helpful to gain as much knowledge of our genetic predispositions as we can. But he has now drawn back from the belief that what we ought to do – how we ought to adapt our sentiments – is just what we are programmed to do anyway. We can, after all, *adapt* sentiments, whether we do so wisely or not. So he undermines his own extreme case and replaces it with the more moderate and reasonable proposal that we should take genetic tendencies into account and not be unrealistic in what we morally demand of people and hold them morally responsible for. It is also important, however, that we do not regard human persons as simply predetermined to do whatever, whether known to them or not, benefits their genes.

3. SELFISH GENES?

The extreme genetic determinist view is sometimes put in the form of a poetic metaphor, the metaphor of the 'selfish gene', ruthlessly seeking its own survival by any means, using humans for its own purposes and dispensing with them without pity. This is a metaphor, because nobody really believes that genes are selfish and have purposes.

From the point of view of evolutionary biology, the selfish gene metaphor is meant to suggest that the basic unit of natural selection is not the species, the group or even the individual organism, but the smallest unit of DNA that is successfully replicated over many generations (cf. Dawkins, *The Selfish Gene*, chapter 3). Such units are 'selfish' because 'any gene which behaves in such a way as to increase its own survival chances in the gene pool at the expense of its alleles [its genetic alternatives] will, by definition, tautologously, tend to survive' (p. 39). This is a controversial issue among biologists, and I am happy to leave discussion of it to them. But two points are perhaps worth mentioning. 'Behaving in such a way as to increase your survival chances' can cover almost any form of behaviour, ranging from slaughtering your rivals to spending all your time helping others. In the right circumstances, both these ways of behaving might increase your survival chances, though it would be odd to call the latter 'selfish' behaviour. The real moral question about selfishness and altruism is what your motivation is – whether you are aiming at your own welfare or at that of others. That question does not, of course, arise in the case of genes, which is why it is so misleading to speak of them as either selfish or altruistic. Dawkins' definition of 'selfishness' is indeed tautologous, as he admits. And it is so vague and wide that, being capable of covering almost any type of behaviour, it fails to isolate one type of behaviour that would ordinarily be called selfish at all. It is also – and this is the second point – morally irrelevant. For it does not follow from the fact that genes behave selfishly (if they do) that human persons, who have been built from genetic recipes, either do or should behave selfishly. There is simply no valid logical path from one assertion to the other. So the discussion of whether genes are 'selfish' or not has no moral relevance to the question of whether morality is an 'illusion'.

Nevertheless, the metaphor of the selfish gene can be used in a highly ideological way, insofar as it can serve to undermine any sense of human distinctiveness, moral agency and purpose. Richard Dawkins, who invented the metaphor of the selfish gene, writes, 'We are survival machines – robot vehicles blindly programmed to preserve the selfish molecules known as genes' (*The Selfish Gene*, p. x). The message this series of metaphors conveys is that humans are machines or robots, that they are instrumental to some deeper purpose (to preserve genes), that they are blindly programmed and that the real business of life is to help genes to survive.

Of course, Dawkins does not really believe this. He says, towards the end of the same book, 'Consciousness is the culmination of an evolutionary trend towards the emancipation of survival machines as executive decision-takers from their ultimate masters, the genes' (p. 63). It is astonishing, but gratifying, to find the inventor of the selfish gene speaking of an 'evolutionary trend' (directionality). Naturally, he wants directionality without a director, but that there is a director is a reasonable supposition, compatible with the evidence we have.

It is equally astonishing to find that Dawkins is not a genetic determinist after all, since humans, however they evolved, have the capacity for rational decision-making. Human rationality and purpose is not an illusion. We can escape the 'tyranny of the selfish genes', those committees of blind midgets from whom it is, after all, possible to escape.

The philosopher Michael Ruse exhibits a similar ambiguity between producing a shocking metaphor that threatens to undermine all morality, and then proceeding to reassure us that we need not worry really. First he produces the shock, saying that in ethics 'we have a collective illusion of the genes' (*Evolutionary Naturalism*, Routledge, New York, 1995, p. 250). Then he gives the reassurance: 'The sociobiologist is committed absolutely and completely to the genuine nature of human altruism' (*Can a Darwinian be a Christian?*, Cambridge University Press, Cambridge, 2001, p. 195).

Just because our moral sense has been genetically predetermined, it does not have to be an illusion. It may be that in the long history of evolution organisms were selected that preferentially propagated specific genes, and that among these genes were some that predispose organisms to exhibit sympathy and altruistic behaviour. This is indeed a helpful explanation of how the human moral sense arose. But it does not at all follow that the moral sense is an illusion. As Daniel Dennett points out (*Darwin's Dangerous*

Idea, Touchstone, New York, 1996, chapter 16), it is an instance of the 'genetic fallacy' to think that once we have shown how a certain capacity arises we have thereby shown its essential nature (I find it pleasing to think that some geneticists fall into their own fallacy). Moral consciousness has arisen through mutation and natural selection, but now that it exists humans are no longer 'survival machines' or 'robot vehicles' for genes. We are free executive decision-makers, moral agents.

In any case, talk of genes being selfish requires some qualification. Far from particular genes trying 'selfishly' to survive, they are constantly changing, producing new and different genes – human beings accumulate about one hundred mutations per generation. Genes, or to be more precise about three per cent of the human genome, have the function of being recipes for proteins, from which organic bodies are built. Among such genes 'the main purpose of most genes in the human genome is regulating the expression of other genes' (Ridley, *Genome*, p. 150). They switch other genes on and off, enabling viruses to be combated, ordering imperfect cells to commit suicide and repairing broken genes. Matt Ridley suggests that about thirty-five per cent of our DNA is truly 'selfish', such genes (if they can be called 'genes') having no function other than replicating themselves. But the function of our most important genes (the vital three per cent) is to build bodies, regulate their development and protect them from internal and external threats. 'We appear to have a general mechanism for suppressing selfish DNA' (p. 130).

Is it really plausible to call this sort of activity 'selfish'? Genes do not survive. They all die. But they copy themselves, they produce variants of themselves and they build bodies and brains. They assemble into collectives (even bacteria are assemblies of about one thousand genes) and must co-operate in exacting ways to succeed in building organisms. Genes are more like selfless recipes (surely a recipe can be selfless if a gene can be selfish) for developing organisms than they are like committees of self-interested egoists.

Even the inventor of the selfish gene qualifies the metaphor by saying, 'The individual organism ... emerges when genes, which at the beginning of evolution were separate, warring entities, gang together in cooperative groups, as "selfish cooperators"' (*Unweaving the Rainbow*, Penguin, Harmondsworth, 1998, p. 308; see also all of chapter 9, 'The Selfish Cooperator'). So even if you were to take the nature of genetic activity as wholly determining moral activity – which is anyway implausible – it is not helpful to move from describing genes as selfish to describing apparent altruism in organisms like ants as 'purely unambiguously selfish' (Ridley, *The Origins of Virtue*, p. 18).

Self-sacrifice by worker ants does help an ant colony to survive. So genes that encourage self-sacrifice are likely to be copied in ant genomes. This is neither real altruism nor real selfishness. It is just what happens, and is morally neutral. Where morality becomes relevant is where an animal species at last develops some ability for self-awareness and self-direction. Then, as heir to a long evolutionary process, it will probably find itself with genes for self-sacrifice as well as for aggression.

It by no means follows that morally chosen altruism is 'really' genetic selfishness. Once such a choice has become possible and they have escaped the tyranny of the genes, humans make a choice to follow some genetic tendencies rather than others. Or, as those who are not committed to atheism might say, they do so when they have achieved that state of conscious self-knowledge towards which the pre-conscious processes of progressive genetic change and environmental adaptiveness have impelled their ancestors. To a believer in God the process looks well designed to produce moral agents progressively from the pre-conscious and simpler physical elements of which the universe is composed.

4. HOMO SAPIENS

So we are led to see *Homo sapiens* as an animal with a complex balance of dispositions, both to compete with and eliminate opponents and also to co-operate and build communities. The first humans were neither perfectly wise and virtuous nor wholly destructive predators. Like us, their descendents, they were a blend of diverse tendencies, destined to ascend from ignorance and pre-linguistic consciousness towards greater knowledge and rationality.

That ascent has proceeded in the last five or six thousand years – a flicker in the history of the cosmos – with remarkable rapidity. Unfortunately, knowledge and rationality have been to a large extent used in the service of selfish and destructive aims. It is from this position that we tend to look back on the process of evolution and see it in terms of warfare, struggle and pointless waste. But it could also be seen as a striving for higher life and forms of consciousness, for greater harmony, complexity and adaptiveness to the cosmic environment, for more intense inwardness, for a conscious creation of new forms of goodness and beauty.

Humans stand at a crucial point in the history of this part of the cosmos, responsible for the further development of life on this planet. So far they have not done too well, but all is not lost. There could yet come to be, as the goal of the evolutionary process, a society of minds that will exhibit intelligence, compassion and bliss, and that will have overcome greed, hatred and ignorance and the suffering and frustration these entail. Whether that is so depends partly on us. Humans have the capacity to choose the good for its own sake.

Why should we choose the good? You may say that the answer is, obviously, just for its own sake. But belief in God can add something to this thought.

The believer in God is one who believes that supreme goodness actually exists, realised in the mind of God. It is not just an invention of the human mind or a purely hypothetical ideal. When humans gain the capacity to choose the good, they gain the capacity to know the Supreme Good, to respond to it and perhaps to be transformed by it. Humans cannot change it, and only if they respond in conscious relationship to it can they realise the deepest fulfilment of human capacities, which is not just to know the good, but to be transformed by goodness itself.

An account of morality that is purely naturalistic explains the sense of obligation as the expression of genetic tendencies imprinted long ago in the history of the race. But if we are genuinely free from the 'tyranny of the genes', we could seek to change these tendencies if they seem disadvantageous. Sociobiologists usually do recommend some 'rational' changes to inherited moral norms. So why should we not change the norms themselves? And if we did, in accordance with what principles would we change them? In accordance with our deepest desires, whatever they happen to be? May we not abandon moral principles themselves if it suits us and if we can?

These are deep waters, and philosophical arguments will continue to rage in them. But anyone who believes in the primacy of consciousness and in the existence of an ultimate mind, the source of all things, is bound to see morality non-naturalistically. The genesis of moral sense will be seen as the genesis of a capacity to know and respond adequately to that reality of supreme value that is the ultimate mind. It is the genesis of a capacity to discriminate truth from falsehood in morality, not just to feel bound by genetically inscribed rules. Like all human cognition, such a capacity will no doubt be liable to many mistakes and will have to develop from ambiguous and uncertain beginnings. But morality will be a process of discovering truths about the nature of goodness, and not a playing-out of routines that were established because once long ago they were evolutionarily stable strategies.

I do not mean that humans can simply discover moral truths by intuiting them as propositions clearly present in the mind of God. Complex moral issues require decisions, analytical thought and sensitive judgment. But such decisions are made in the light of our deepest beliefs about the nature of reality and of human nature. To the extent that humans gain a deeper understanding of the nature and purpose of God they will be better placed to make such decisions with true wisdom. For that which is morally right will, for the theist, be identical with that which is in accordance with the purposes of God. And to know that is to know something about the nature of God.

Most importantly, those who believe that morality is grounded in the being and purposes of God will feel, or certainly should feel, that there is a deeper reality always to be further understood, a morally objective truth to be found, and a vision of goodness that is only at present dimly discerned. These things give a seriousness and significance to morality, but also suggest a degree of humility about whether we have yet truly discerned what is objectively right. This is dimensionally different from a belief that morality is the imprinted echo of what made for survival in the prehistory of the human species.

To me, it makes sense of the long evolutionary climb towards self-consciousness and moral responsibility to say that the ascent to the vision and love of the good and beautiful is its goal, or at least one of its major goals. When that goal is attained, humans become capable of conscious relationship to a Supreme Good. It is in such a transforming relationship that morality is fulfilled and the ultimate meaning of human existence is to be found.

5. COSMIC ONTOGENESIS

The evolutionary story can be taken as a demonstration that mental and personal life is at base purely material and requires no reference to any spiritual reality. The human soul is not inserted into the material body as a completely different sort – a spiritual sort – of existent. The soul – if it even makes sense to speak of it as though it were a distinct entity – develops as a complex material entity by millions of small steps from the interactions of simple physical elements. So it looks as though the 'greater' (the personal) can and does evolve naturally and without supernatural help from the 'lesser' (the physical). The classical argument that mind cannot emerge out of matter, that the greater cannot come from the lesser, is refuted once and for all by the facts of evolution. Materialism might seem to be vindicated by a realisation of the continuum of natural evolution.

There is another way of reading the story, however. After all, the problem of consciousness has not disappeared. However we twist and turn, the obstinately resistant fact is that personal experience, with its interiority and subjective feeling inaccessible to others, does seem different in kind from the observable physical brain states that are undoubtedly the basis of human consciousness.

The incredibly organised complexity that is required for the sub-atomic particles of the material world to form stable elements and self-replicating molecules, central nervous systems and brains – all of which form the basis for consciousness – is awe inspiring. Though it *could* all occur by chance, to many people it is almost irresistably suggestive of design, and design of such elegance and complexity that human intelligence cannot yet replicate it.

So it can seem that evolutionary history was meant to realise a form of being that is unique and distinctive, the transformation of physical reality into personal reality. In biology the term 'ontogenesis' is used to denote the development of an organism from the earliest embryonic stage to maturity. The Roman Catholic palaeontologist and visionary Pierre Teilhard de Chardin spoke of a cosmic ontogenesis to denote the development of the cosmos from its earliest physical stages to its maturity as a fully personal reality (*The Phenomenon of Man*).

From this viewpoint, it was no accident that the complicated way in which the universe expanded immediately after the big bang led to the formation of the particles, atoms and elements that would make organic life possible. These particles, and the systems they constituted, always had the interiority that would lead, in the course of their natural ontogenesis over millions of years, to consciousness. Conscious, self-aware, morally responsible persons are the mature development of what was embryonic in the earliest stages of the physical universe. Human persons are neither supernatural intrusions into a purely physical universe nor bits of matter shuffling purposelessly about. They are natural developments of a cosmos we call material but whose inner nature is shown by quantum physics to be more hidden, complex and mysterious than materialism ever suspected. In seeking an ultimate explanation of the universe, we may need to take account of such ontogenesis. It suggests that there is a proper goal of cosmic evolution, a goal that the universe will necessarily make possible in some way. If there is a goal, it is reasonable to suppose that there is an ultimate self-existent mind that sets the goal and ensures its real-isation. And it is plausible to think that human persons are at least part of that goal, points at which, if only fitfully and in part, the cosmos realises its inmost potentiality. Humans may not be the final term of cosmic ontogenesis, but they are points at which the cosmos realises its potentiality of personal being. Minds do evolve from matter; but matter was always destined to realise its potentiality in mind.

6. THE GOD OF THE PHILOSOPHERS

In part 2, I have raised the question of what an ultimate explanation of the existence of the cosmos might be. I have argued that there are many aspects of reality, especially aspects connected with mind, consciousness, purpose and value, with which the natural sciences do not deal. Any adequate explanation will have to include these aspects, and so be consilient in the sense of har-monising personal and scientific explanations in one coherent whole.

I have proposed that the crucial element in such an explan-ation would be the existence of an ultimate mind, the necessarily actual and self-existent basis of all possibilities, actualising some set of possibilities for the sake of the distinctive sorts of goodness they exhibit. In this cosmos the history of the emergence of con-sciousness and creative responsible freedom from what appears to be a simple material foundation articulates the sorts of good-ness that cosmic evolution produces, and the sorts of limitations on goodness that are its consequences. The fact of cosmic evolu-tion can reasonably be seen to point to the possible goal of the cosmos as the emergence of a society of intelligent, morally free and potentially compassionate minds. That in turn points to the existence of one ultimate mind, one supreme intelligence, that is the origin and director of the process.

All this, however, might seem to point to 'the God of the philosophers and men of science' rather than to the living 'God of Abraham, the God of Isaac, the God of Jacob', as Pascal put it

(Blaise Pascal, 'The Memorial', 1654, found sewn into his clothing at his death). A properly informed scientific view of the cosmos might well be sympathetic to the idea of an originating mind of vast intelligence, even of supreme perfection. But what place is there in all this for the God of the Bible and the Qur'an, of the *Bhagavad Gita* and the *Guru Granth Sahib*?

The scientists' God seems a remote and rather unconcerned mind, who does not act to perform miracles, to interrupt the laws of nature, to answer prayers and to make specific and often peculiar demands on believers (like what sorts of clothes they should wear or what sorts of food they should eat). The scientists' God is rather like Aristotle's, a God who remains happy and unmoved by all that happens in the world, though the process of cosmic history is set up and directed by God in a general way to achieve valuable goals. How, though, does all this connect with the religious God who acts in specific ways, who requires people to worship and obey and who makes specific promises to true believers?

Specific religious beliefs about God depend on some sort of revelation from God. The God of science, naturally enough, does not mention revelation at all. So there are obviously going to be some differences between the God of science and the God of religion. Yet is it possible that there are features of the scientific quest for ultimate explanation that actually point towards a more religious idea of God? Can science even lead us to expect some sort of revelation?

In part 3 I will explore this possibility. My main argument is that the religious and scientific Gods are not as far apart as we might think. I am not going to defend a specific religious view, but I will suggest that, in a general way, commitment to an intelligent creator might lead us to expect some further non-science-based information about, and even personal interaction with, this creator. It might lead us to seek further information about the final purpose of the cosmos, beyond the rather speculative conjectures that contemporary scientists sometimes hazard. If there is such information, it will go well beyond the limits of science, but, rather surprisingly, it turns out that science has some very interesting things to say about such further possibilities.

PART III

THE GOD OF RELIGION

14

GOD'S ACTION IN
THE UNIVERSE

God's intentions for the universe will influence how things go, perhaps by a 'whole–part' form of causation. But there may also be specific divine acts in the cosmos. These will usually work within the general scope of the laws of nature and can be construed as divine influences on the possible tracks into an open future that the physical nexus provides. Prayer can be conceived as providing influences from intelligent agents in the cosmos that affect the specific determination of the open future. Thus ultimate mind will both possess a form (possibly many forms) of temporality and also exist beyond all times.

AIM: To show how God can act in a law-governed universe. To show how God may possess a form of temporality, so that God can be influenced by prayer, and may act in specific ways to influence a partly open future.

1. COMPASSIONATE MIND?

The natural sciences have brilliantly vindicated the method of discovering the nature of reality by repeated publicly accessible observation, precise measurement and the testing of bold causal hypotheses by experimental verification. This method, however, has no access to the world of mental experience, where thoughts, feelings and intentions cannot be publicly observed, precisely measured or verified and where causal influence is often in terms of the striving for new and unique goals rather than in terms of repeatable and quantifiable regularities. The mind typically proceeds by visualisation, evaluation and goal-oriented action, not by obedience to impersonal laws.

If there is an ultimate mind, the basis of all physical reality, it follows that the natural sciences have no direct access to it, nor to the formation of divine goals and their implementation in physical contexts. The sciences can, however, provide data that may help us to decide whether there may be such divine goals and intentional actions in the cosmos, and if so what their general character seems to be, in the light of the facts that the sciences disclose.

My argument has been that the cosmos revealed by the sciences can reasonably be seen as originated by a divine mind. The sciences do not suggest, however, that this mind interacts with humans in a personal way, communicating information about the divine purposes and responding to human requests or thoughts. It rather exists as a supreme intelligence from which all worlds flow as manifestations of its infinite potency and fecundity, and whose ultimate goals far transcend all human concerns.

Yet I have argued that the ultimate mind will choose for itself a set of states of supreme beauty and goodness. We might expect it to be supremely happy in the creation and contemplation of many forms of intrinsic values. It may also create a universe for the sake of the goodness that the universe actualises. It will, we may suppose, enjoy those created forms of goodness. By the same token, it may react with something like sadness or sympathy to the sufferings, frustrations, conflicts and disasters to which organic life is prone.

Though science cannot assure us of this fact, it seems plausible to think that any being of supreme goodness would not be indifferent to the sufferings and joys of humanity. If there is an ultimate mind of supreme goodness, in a world in which suffering plays an important part, it would not be unreasonable to think that it would be a supremely compassionate mind. Humans, in their capacity to create and appreciate beauty and goodness, are images, however faint and corrupted, of that supreme intelligence. So there is some reason to think, or at least hope, that a cosmic mind might in some way help humans achieve their small but divinely intended goals, and alleviate or transform, so far as is possible, their sufferings.

How and in what ways God might do this is beyond the province of the sciences. The various religions of the world depict ways in which the human experience of suffering, failure and slavery to egoistic desire may be overcome, and in which a reality of supreme goodness might help the human quest for liberation from evil and perhaps even ultimately liberate creatures from evil. Nevertheless, the doctrines of most world religions were formulated before the scientific revolution, and the picture of the world given by science suggests revisions in many traditional religious views of the way in which a divine mind is likely to act.

2. DIVINE ACTS AND LAWS OF NATURE

The most important revision is the replacement of a worldview in which God or lesser spirits directly and immediately cause specific events in response to human acts or wishes, by a worldview in which the universe develops largely by its own inherent principles in accordance with general laws of nature. Some would say that the scientific view rules out any acts of a spiritual being at all. But from Isaac Newton onwards it has been fairly clear that this is not so. If there is a God who creates the laws of nature, that God could also act in ways not falling under laws of nature. In fact the creation of the laws of nature itself cannot be in accordance with any laws of nature, so the possibility of divine action beyond the laws of nature is established by the initial hypothesis of creation.

The crucial feature of personal action, in contrast to purely physical causation, is that events are caused to happen in accordance with, and because of, the formation of an intention. In our universe humans often cause things to happen by forming an idea of a goal and then acting to achieve it. I have construed this as a development of more primitive responses of organisms to sensory stimuli. But, at least for common sense, causation by intention introduces a new causal factor into the physical situation. That factor would not be present where consciousness and intelligent agency did not exist – namely, at less developed stages of evolution.

At least it would not be present in the physical world. But if the whole physical world is generated by intelligent agency, then divine consciousness and intention would be present at every moment of the universe's existence, though not as a property of physical organisms. The purposes of the cosmic mind will always be present and might be expected to have an influence upon the way the universe develops.

The character of such divine influence would, we might suppose, be compatible with the existence of the general laws of nature, with the general structure of chance, freedom and necessity that characterises the universe, and with those features of entanglement, emergence and self-organisation that are suggested by modern science.

What this suggests is that any divine action is likely to be co-operative, directive or persuasive rather than wholly determining or interruptive of natural processes. It may be the sort of 'whole–part' causation of which Arthur Peacocke writes (*Paths from Science towards God*, chapter 8.3). Intentions in the divine mind will exercise a general influence on the way the cosmos develops, without actually interfering with any of the laws of nature. Think of the way in which a person's intention to write a book may exist over many years. It will guide many of that person's actions and thoughts, but it will not normally introduce direct supernatural causes into the writing process. That a person intends to write a book will certainly guide such things as their reading, their use of leisure time and their acquisition of writing skills. The person will do things they otherwise would not have done. But there will be no obvious non-physical events interrupting the normal causal processes of that person's life, no

violations of laws of nature and no cancellation of normal capacities and behaviour patterns. So we might think of God's intentions as guiding and shaping natural processes, especially if there
are crucial points where different tracks into the future are open,
without interrupting or violating the course of nature.

3. CONSTRAINTS ON DIVINE ACTION

There is no problem about a divine mind directly causing any
event it chooses. Most, if not all, such divine acts would be scientifically undetectable. They may occur unobserved, or in areas
where measurement and repeated experiment are impossible, and
they would not affect the general structure of the physical world.
They may be undetectable in principle by any physical methods.
In quantum physics probability waves carry no energy, but give
probabilistic direction to the momentum of particles. By analogy
we might think of a divine causality without physical energy that
gives direction to diverse possible tracks into the future, influencing in a stochastic or statistical way the creative causality of finite
objects.

The problem is not of the possibility of divine action in general. It is the problem of what the constraints on divine action are,
and of why it should accept such constraints. Science can help to
outline some of those constraints and suggest reasons for their
acceptance.

One reason for accepting a law-governed universe is that it
allows different subjects to form and express their desires in a
shared common space. There is a recalcitrant reality to be shared,
struggled with, overcome and shaped. No coherent pattern of
desires could be established without a publicly shared arena in
which they can grow. Humans are beings that can understand,
appreciate and shape the world for good or ill. There must be
resistant objects to be understood, and there must be a neutral
world if it is truly to be a common environment, shared by very
different persons.

Human existence is a striving, a self-development that
requires an objective 'other' to realise itself, that involves shared

experiences and actions to achieve wider understanding and forms of creativity. It is a social self-realisation that is part of the emergent process of the universe, as human existence takes responsibility for its own future. The laws of such a universe must be morally neutral, posing challenges as well as opportunities that are roughly the same for all, and in which good and bad are inextricably intertwined in one reality that strives towards realisation of its own inner potentialities.

Humans live in communities that enable experiences to be complemented, corrected and refined by others. Forms of creativity are possible in co-operation with others that would be impossible on our own (it is impossible to play a violin concerto without an orchestra). Thus the existence of a physical world with morally neutral general principles of causality (the laws of nature) is necessary to the development of the sort of personal life that human beings have.

God, the primary consciousness, may play an integral part in this process, for humans may be sensitive, probably at a largely subconscious level, to divine reality in its intrinsic perfection and in its envisaging of the ideals towards which the processes of nature should be moving. Such sensitivity, where it occurs, may be able to correct and re-align our consciousness of our own future possibilities.

4. THE FLOW OF TIME

So far the divine intentions have been construed in a timeless way, as though God's intentions never change but exist in the divine mind unchangeably and completely beyond time. Finite minds, as they pass through time, may come to know something of those intentions, but the intentions themselves will never change. This is how most traditional religious thinkers have thought of God, as a timeless being beyond the flow of time.

Many scientists are happy with the idea of a wholly transcendent changeless mind, beyond both space and time. One reason they like the idea of a mind completely beyond time is that

Newtonian physics is time indifferent. It does not matter whether you run the laws of mechanics forwards or backwards. Relativity physics is also comfortable with a 'block time' view – the view that all of time exists at once in a four or more dimensional continuum. Time becomes a fourth co-ordinate along with the three dimensions of space, and perhaps with other co-ordinates too – M theory, a modern theory in mathematical physics, posits that there are eleven dimensions, 'rolled up' and invisible in ordinary human experience.

This is very difficult to envisage, and it is not necessary for most of us to try. But many of us are familiar with the representation of four-dimensional space-time as a sort of grid, on which the space-time locations of everything can be plotted as 'world lines', lines running through a number of points in space-time. All human lives can be represented as lines in space-time. We seem to move along those lines, but in fact many physicists think of all human lives, and all events in space-time, not as moving but as existing timelessly (cf. *God, Time and the Creation of the Universe*, Chris Isham and Keith Ward, Royal Society of Arts, London, 1993, pp. 55–69).

Physicist Max Tegmark puts the view with startling frankness: 'If parallel universes contain all possible arrangements of matter, then time is simply a way to put those universes into a sequence. The universes themselves are static, change is an illusion, albeit an interesting one' ('Parallel Universes', *Scientific American*, May 2003, p. 47).

Our futures timelessly exist in space-time, though they seem as yet non-existent to us. In that sense, many classical and relativity physicists regard the flow of time, the sense that it is passing from a completed past to an open future, as an illusion. Such views can easily work with the idea of a timeless God.

In some modern science, however, there has been a suggestion that the flow of time is a fundamental feature of reality, not reducible to just another co-ordinate on a completed grid. The second law of thermodynamics gives time an irreversible direction, from lower to higher entropy. And if nature is not deterministic, there may be alternative tracks into an open future.

Nobel laureate Ilya Prigogine and Isabelle Stengers, in the final chapter of their book *Order out of Chaos* (Heinemann, London, 1984), based on Prigogine's work on unstable dynamical systems far from equilibrium, suggest that time is not, as classical physicists sometimes thought, an irrelevant factor in the operation of physical laws. Rather, as they put it, 'The arrow of time is the manifestation of the fact that the future is not given' (p. 16). Again, 'Randomness and irreversibility play an ever-increasing role. Science is rediscovering time' (p. xxviii).

In nature, they suggest, there is genuine becoming and room for free creativity and the sort of directionality that may well suggest purpose. There is sensitivity to context and a seemingly non-deterministic, and certainly non-predictable, character to basic physical processes that suggests they have a more holistic and open character than classical physics allowed. Thus nature may be less like a deterministic machine than it is like an emergent and dynamic process, with many possibilities of development and creative novelty.

Arthur Peacocke and John Haught similarly suggest that the evolutionary process is one of the emergence of complexity, 'characterised by propensities towards increase in complexity, information processing and storage, consciousness, sensitivity to pain, and even self-consciousness' (Peacocke, 'Biological Evolution', in *Evolutionary and Molecular Biology*, ed. Robert Russell, William Stoeger and Francisco Ayala, Vatican Observatory, Vatican City, 1998, p. 367). It is as if there is an insistent drive towards novelty and complexity, but not divine pre-determination of every specific event. Haught suggests that this implies a vulnerable, loving, participatory God, a God who is self-giving, persuasive to emergence, protective of autonomy and oriented to the future ('Darwin's Gift to Theology', in *Evolutionary and Molecular Biology*, pp. 393–401). From an evolutionary perspective, God may be seen not so much as an all-determining sovereign, but as the ideal that invites, the energy that drives, and the mathematical wisdom that orders the cosmos.

For such views, the cosmos is a dynamic, creative, emergent totality. The future is not predetermined and already existent, a block through which we seem to travel, while in fact our lives are changelessly existent from beginning to end. Rather, the future is decreed in its general outlines by the creative mind from which it is generated, but it is open to the truly creative and original choices of many finite agents that will help to determine the paths it takes towards and the detailed specification of the general goals of creation.

5. THE OPEN FUTURE

This is another point at which a decision needs to be made between a deterministic and a more open, creative view of time. Insofar as we regard the space-time grids of physics as useful symbolic abstractions, and take mental life to be the fundamental reality, we might incline to an 'open future' view. For time is a characteristic of most mental life as it flows from one feeling and idea to another. As Aristotle said, time is the 'measure of change'. There cannot be change without time, but if it is possible for conscious states to exist without any physical states (in God, for example) there can be mental change without physical change. There can logically be thoughts that succeed one another, creative decisions and developing feelings without any observable bodily changes. Is there such change in the life of ultimate mind, in the mind of God?

Most classical religions have decided that God is changeless and therefore timeless, since perfection is changeless, and the creator of time cannot itself be bound by time. The consequence of this view is that God must create the universe in one changeless act. God cannot wait until some things have happened before deciding what to do next. The whole of time, from first to last, must be created in the same act (or 'at the same time'). No changes in the divine plan are possible. (For a full treatment of this theme, see Keith Ward, *Religion and Creation*, Clarendon Press, Oxford, 1996, chapter 11.)

Suppose, however, that finite agents have the sort of radical autonomy that consists in their not being wholly determined either by God or by any previous state of the universe. Then God would not, by one non-temporal act, make everything to be what it is, since finite agents would partly determine the future. God's knowledge of what is the case will be partly dependent on what finite agents have determined in the universe, and what God creates next will partly depend upon that knowledge. This will entail that there is temporality in God, since some of God's knowledge of and God's particular intentions for the universe will logically be subsequent to ('after') the occurrence of some contingent states of the universe, which God has not determined. The divine plan for the universe cannot be completed all at once. Some of it must wait for decisions made in the course of the history of the universe. That means that there must be different successive states of the divine being, and that is possible only where time exists.

If, as I have argued, radical freedom is a condition of real moral responsibility and human autonomy, then autonomy only truly exists if there are alternative possibilities of action, undetermined by any prior internal or external necessity. Temporality is a condition of this sort of contingency – if there are alternative possibilities, then there must be a time when possibilities are open, followed by a time when one of those possibilities has become actual. So time is a necessary condition of the sort of change from potentiality to actuality that constitutes autonomy and radical freedom. If there is real autonomy in the universe, the divine mind must be temporal in at least one aspect of its being. Divine temporality is not a concession to the sentimental wish for a caring personal God. It is a necessary condition of true divine knowledge of a world in temporal flux, of true finite freedom and of true divine creativity.

6. DIVINE MEMORY

This may lead one to suspect that temporality is actually a perfection, something to be positively valued. At least it is a necessary condition of something of great value, the sensitivity and creativity of personal being. These may well be attributes a perfect being would possess.

Many classical philosophers and theologians, and many modern physicists, have supposed that time is imperfect. It involves perpetual loss, as all things pass into nothingness as though they had never been. So many good things lost, it may seem, so many efforts and achievements, and all of them to be forgotten, never to be recalled. How much more preferable is the unchanging world of eternal essences or mathematical truths, for ever beyond the vagaries of time and decay. Even Bertrand Russell, that great atheist, said that in pure mathematics he came nearest to a truly mystical experience of a pure world, deathless and beyond the grasp of time.

Yet much modern physics points to the importance of the temporal, the particular, the individual as the key to reality. Quantum physics, which is above all the realm of mathematical beauty, points to the importance of time in resolving probabilities into observed actualities; if it points to a veiled Platonic reality underlying the world of appearance, it stresses in an emphatic way the incompleteness of intelligible potentialities without some phenomenal (spatio-temporal) actualisation.

Temporal actualisation adds something to eternal possibility. So it may be that, if the divine mind is fully real, it may have to be temporal. In it, however, there is no loss or forgetfulness. If the divine mind knows everything it is possible for any being to know, then whatever is once actual will never be lost. It will be remembered for ever. Moreover, in memory the immediacy of experience can be transformed by its incorporation into a wider and ever-expanding consciousness. The unbearable pain of some experiences can be mitigated without being destroyed, as time lessens their impact, and their intensity is lessened. This is one way in which suffering can be transformed – not by being turned into good, but by being pushed to the edges of full awareness while being accepted as part of an experience that is overwhelmingly good. Only time and the refocusing of attention that memory permits can have this healing and transformative function. It could be part of the perfection of the divine mind that it can preserve every experience without loss, while continually re-ordering the intensity of past experiences so that the focus of attention is on the good, and the bad can be more dispassionately seen as a necessary but transformable part of the totality of experience. That is only possible if there is temporality, change and memory in the divine mind.

7. DIVINE TEMPORALITY

In a similar way the transience of the present can, in certain moods, seem an imperfection, since nothing seems stable and enduring. Yet without transience there can never be any new experiences; there can never be the possibility of doing something new and for the first time, of real creative originality. If it is an excellence of mind to be creative, then there must, even for ultimate mind, be the opportunity for creatively bringing new things into being. Aristotle defined true happiness as lying in the unimpeded exercise of the proper capacities of a rational being. In the post-sixteenth-century scientific world we regard creative insight and understanding as a very great value, and so perhaps we see more clearly that the ultimately wise divine mind may realise its perfection in the imaginative genesis of new forms of beauty and beatitude. True creativity presupposes the existence of time.

The openness of the future, too, can be seen as a perfection and not an imperfection in God. For it does not mean, as it does so often for us, that God has no idea what will happen next, or that something bad may happen unexpectedly to God. The future remains always under the control of the divine mind and is no threat to it. But such openness allows the space within which the divine mind can enter into real relationship with the finite minds that the cosmos generates from itself.

Finite subjects have the capacity to appreciate and to celebrate difference, to share their unique and diverse experiences and to co-operate in creative activities, so that some of their greatest goods are found in living and growing together. Human persons, in their ambiguous world, so often bound by selfish desire, can help one another to develop their own unique capacities, to live through the obstacles and sufferings that they encounter and to encourage one another in loyalty and friendship to face the future with courage and hope. They can even, when called upon to do so, give their own lives so that others might live more fully.

The divine mind will know, and will surely feel, all these things. It will know human feelings and experiences as if from the inside, in an intimate and intense way. It will respect the autonomy of human actions and feel compassion with humans in their sufferings. Yet a mind that watches others walking into disaster, a compassion that is without corresponding action, seems poor and impoverished, even morally offensive. It seems a greater perfection to respond actively, in whatever ways are appropriate, to save humans from the worst consequences of their often absurd choices, and to encourage them to live happier and more fulfilled lives.

This means a continuing adjustment of possibilities, of discouraging and encouraging factors in the human world, as circumstances change. The mind of the cosmos will not only be in itself supremely happy. In a world that lies under the sovereignty of disordered desire, it will also be a mind of supreme compassion.

It will not only feel the sufferings of others. It will give, to whatever extent is possible, some awareness of the wisdom and bliss and beauty of the divine mind to those seek it. It will not only observe the actions of others. It will seek to co-operate with them, insofar as they allow it, by enhancing their capacities for goodness. It will, to the extent such a thing is possible, renounce its own supreme bliss to share in the imperfect world of creatures, by sharing their experiences and co-operating with their actions, if by so doing it could begin to transform that imperfection.

All these things involve a temporal and continuing responsiveness to the experiences of finite subjects. Just as the past is continually re-ordered in the divine mind to heal the memory of suffering, so possibilities for the future are continually re-ordered to open up new paths to liberation from selfish desire. So in these ways – the healing of the past, the exercise of creativity in every present, and responsiveness to the free acts of creatures that continually re-order possibilities for the future – temporality might well be a perfection of ultimate mind.

8. PRAYER

The objective reality of the divine mind, and of the goals and ideals within it, is the causal basis of the whole universe. If the universe moves towards an open future, the divine mind will exercise a continuing causal influence on the way things go. It will impede processes that frustrate the realisation of divinely willed goals (in religious terms, this would be 'judgement'). It will aid processes that move towards such realisation (this would be 'providence'). It will co-operate with finite minds that are open to its influence (these would be called moments of 'inspiration and healing').

None of this will 'interfere' with nature, as some atheists put it. It is more a matter of working with the realisation of natural powers. We might imagine that, just as the organisation of a living body switches on particular capacities of cells, all of which originally had the potential to grow in different ways, so the reality of ultimate mind switches on and encourages the development of various natural capacities and discourages others. When finite minds come into being, they are free to be open to divine reality or to be closed to it because of egoistic desire. Prayer can be seen fundamentally as the opening of finite mind to ultimate mind. The expression of disinterested desire and a will for objective goodness will be among the factors to which ultimate mind responds as it shapes the future. Thus prayers will have a real effect on what happens. Yet often the real prayer will not be the one people utter, but the hidden desire that is in their hearts, and that desire may be one that any truly good mind would have to oppose. In addition, in a universe of entangled connectedness, law-like necessity, freedom and self-shaping, the responsive actions of even a supremely powerful mind will work with the developing potentialities of things, and so will take into account innumerable factors of which we can have no knowledge.

This suggests that prayer will be effective in a mind-based cosmos, but not in the sense that whatever we ask for will happen – such a thing would be chaotic, disastrous and subversive of the possibility of an intelligible and self-organising universe. Rather, the hopes and intentions of finite minds will be known to infinite mind and will be taken into account as causal factors in shaping a partly open future.

9. GOD BEYOND TIME

It should not be thought, however, that God is temporal in the sense of being bound to our time, as if the whole of the divine life is constrained to go along moment by moment with us. Such a thought would raise in a very acute form a problem with the special theory of relativity. The problem is that in our space-time simultaneity is relative. That is, with respect to two observers moving relative to one another, two events that are simultaneous for one observer may occur one after the other for the other observer. There is no such thing as absolute simultaneity throughout the universe. Where, then, is God's 'now' if there is no absolute flow of time which God could observe or share in?

Two things should immediately be said. One is that the local ordering of events in time is invariant. Whatever things look like to different observers, the order in which events relate locally to one another is fixed. There is, in other words, an absolute causal sequence between local events, which special relativity does not undermine.

The other is that, since our cosmos originated with the big bang, and all events in the cosmos have occurred since then, there is an absolute ordering of time with reference to the relation of each event to the big bang itself. Special relativity just shows that there is no absolute and independently existing time through which events flow. But there is no problem for a God who can directly relate in knowledge to the local causal ordering of all events, without having to relate all events in the cosmos to one absolute present.

Indeed God can go much further than being aware of two different processes of temporal succession that cannot be related to one another by some absolute measure of simultaneity. As many modern cosmologists postulate, God can create two or more space-time universes that are not temporally related to one another at all ('parallel universes'). So there will be no 'now' at which events in these universes can be correlated. God will know each time successively, as it flows, but will not relate those successions to one another temporally. Thus God is not to be temporally located in one of those time-streams. God simply sees them both as they are – successive. There could be a great number of such universes, and God would apprehend the successivity of each.

God can enter into many different times, acting and responding in them, while also existing in a trans-temporal way. We cannot imagine this trans-temporality of God, but it should not be conceived as a totally immutable and static existence. It might be better conceived as a transcendent agency that acts incessantly in many temporal streams, manifesting its changeless perfection in continual creative activity, sensitive awareness and overflowing goodness.

There are respects in which God is timeless. God's reality transcends any created time and is not bounded by any. Yet God is actively creative in every temporal universe there is. God's essential trans-temporal nature is wholly unimaginable by us, except that we can say, using the best terms we have from our experience and projecting them towards an inconceivable fullness, that the divine nature is perfect beauty, compassion, wisdom and bliss. But in relation to this and any universe we can imagine, these attributes will be actualised in particular acts and manifestations, some of which can be apprehended by the sciences, and some of which can only be discerned in personal and historical experience.

15

BEYOND THE POWERS
OF NATURE?

Religious faith depends upon experiential, not physical, evidence and requires total practical commitment in objective uncertainty – as many human decisions do. The occurrence of miracles, events transcending the ordinary powers of nature, is rather probable, given the existence of an ultimate mind that wills its nature and purpose to be known. The arguments of some philosophers that science disallows miracles, or that no evidence could ever justify belief in miracles, are flawed.

AIM: To show that faith in God can be reasonable and give rise to absolute commitment, even when such faith is theoretically uncertain, in the strict sense of being based on publicly accessible and experimentally repeatable knowledge. To show that miracles are possible and even rather likely, if there is a God, and that God might be expected to act in extraordinary ways to reveal the divine nature and purpose.

1. THE SOURCES OF RELIGIOUS BELIEF

I have argued that the sciences show the plausibility of the postulate of an ultimate mind, the source of all and the supremely perfect form of reality. But even at that level there are those who see the sciences as having a very different story to tell, a story of a vast cosmos without purpose or meaning, in which mind is only a fleeting by-product of blind impersonal processes. As I have gone on to spell out what it means for there to be a cosmic mind of supreme beauty and compassion, I have had to make a number of choices that are also contested. I have chosen for human autonomy and radical freedom, for the fundamental and irreducible reality of time and for the unique value of personal experience. I believe these are all choices made plausible by important strands of modern science, though there are other strands that point towards different choices. Although what I have said is strongly influenced by modern science, it is also provisional and to some degree tentative. I am not theoretically certain that I am correct. But I am in practice wholly committed to it.

How is this possible? Because in the end these decisions depend upon our own personal experiences of life and our evaluation of them. If I think there is no evidence in human history for the existence and activity of ultimate mind, or if I rate such evidence as there is as inadequate or untrustworthy, I may not be much impressed by theoretical arguments about it. If there is evidence, the discussion so far suggests that it is likely to lie in experiences and events connected with the lives of persons who are especially open to the presence and activity of God. Such persons are likely to be rare in our world, but in them a sense of the presence and the character of God may be strong and enduring, and their natural human powers may be heightened because of their ability to mediate the divine life more fully.

The source of a vital belief in God as divine mind is likely to lie, for those who have it, in some personal experience of divine presence and of divine guidance or empowerment. It will lie in a natural sense of the personal character of reality. For some, science's disclosure of a universe ordered by impersonal laws and of the vast and apparently empty darkness between the stars counteracts this natural sense of the personal. But, as I have shown, for others, and for many of the most original scientists, this disclosure serves to amplify a sense of the intelligibility and grandeur of God.

Modern science suggests a specific view of what God is likely to be and do in a universe such as this. But it does not of itself convey that sense of personal encounter, whether in the beauty of the natural world or in the hidden experiences of the human heart, that turns a tentative theoretical belief in God into a passionate striving for liberation from hatred, greed and ignorance and for a deeper and closer unity with the beauty, wisdom and compassion of ultimate mind.

Such a passionate striving for unity with ultimate mind is the core of religion, and I think it would be wise to look for guidance to those who have achieved, or who have come close to achieving, such liberation and unity and who show it by the purity of their lives, the wisdom of their teaching and the extraordinary effectiveness of their actions in liberating others for lives centred on goodness.

2. EXPERIENTIAL EVIDENCE

Evidence is important in matters of religion. But it is not the sort of evidence with which the sciences deal – publicly observable, experimentally repeatable and quantitatively measurable. It is the sort of evidence upon which we base our daily lives, our relationships with other people and our fundamental decisions about how we are going to live.

In such matters, personal evaluation is inescapable, universal agreement is unobtainable, and seeking to achieve theoretical certainty is inappropriate. We have to make commitments, sometimes with huge consequences, by balancing such things as our own trust and suspicion, rashness and indecisiveness, compassion and judgement. We have to decide such things as what is likely to bring us happiness, what we are likely to be good at and to whom or what we should give our loyalty. We do so by drawing together as wide a class of evidence as we can – but that evidence consists of such insubstantial things as memories and evaluations of past experiences, belief or disbelief in the testimonies of others, general judgements about human character and history, and personal intuitions about what things are right and wrong. All this is inescapably personal and subjective in the sense that it depends to a great degree upon perspectives and attitudes that belong to this place and time and no other. We may, and we no doubt should, try to make such judgements as objective – as reasoned and dispassionate – as possible. But in the end the conclusion we draw will be a decision that we make in the knowledge that others might decide differently. The drawing together of evidence to a decision is a matter of the way we connect the data, the relative importance we give to various data, the perspective from which we see the data, and the patterns we see in the data.

The Cambridge philosopher John Wisdom called this the 'connecting technique' (see his profound and influential paper 'Gods', in *Logic and Language*, 1st series, ed. Antony Flew, Blackwell, Oxford, 1963). This is more like the way in which a judge sums up the evidence in a criminal trial than like the application of some standardised routine. It is not a matter of valid or invalid deduction or inference. It is a matter of synoptic judgement on a range of disparate evidence, all subject to personal evaluation and incapable of being viewed with complete dispassion. It is often, as Kierkegaard said, a matter of passionate commitment made in a situation of objective uncertainty (Søren Kierkegaard, *Concluding Unscientific Postscript*, ed. and trans. H.V. and E.H. Hong, Princeton University Press, Princeton, 1992, p. 203).

This is not pointing to a disadvantage of experiential evidence compared with the greater certainties of science. It is just to point out what experiential evidence is and a reminder that most fundamental real life decisions are made on the basis of non-scientific evidence of this sort. So in matters of religion, when we consider our own relationships with religious believers, the sorts of experience we ourselves associate with religious belief (perhaps we have seen it as naïve and superstitious, or perhaps we associate it with some of the most powerful and beneficial feelings we have had), and our judgements of the effects of religion in history, we will come to a personal evaluation of religious belief.

For some, most religious believers will be hypocrites or censorious dimwits. Religious belief will be blind dogmatism, opposed to the careful truth testing of science. And religion will be seen as the cause of war and violence throughout history. It will then be highly unlikely that any incipient experience of the cosmos as a medium of personal encounter will be welcomed or seriously considered.

For others, however, some religious believers will have been found to be heroically moral and profoundly wise. Religious feelings of awe, dependence and awareness of a personal presence will be valued as among the most important in life. And religion will be seen as productive of the greatest art, the greatest virtue and the deepest forms of understanding, including the scientific understanding of the cosmos. For them, religion will be the fulfilment of the most distinctively human experiences and aspirations, and the evidence of personal experience of God's presence and power, in oneself and in others, will seem strong and convincing.

Experiential evidence is fundamentally important to everyone. But it is always likely to be contested. Since matters of religion are matters of one's most basic attitudes to existence, the contest over the truth and plausibility of religion is likely to be most basic and profound. That is why evidence is relevant to religious belief but is incapable of settling the issue of truth in this area. Therefore, our duty is to comprehend the nature of evidence in this area, to get the best evidence we can, to try to ensure that we put things as fairly as we can, to evaluate the evidence as justly as we can, but not to look for universal assent.

3. SAINTS AND PROPHETS

Thus in religion we are primarily looking for experiential and contested evidence, not for experimentally testable and agreed evidence. Those who accept the postulate of God might expect to find evidence of divine inspiration and empowerment, though rather rarely and focused on persons who are unusually attuned to the divine mind, who are, in religious terms, 'close to God'. Such people might show extraordinary moral resolution, display amazing insight into human motivations and problems, have an intimate and infectious sense of God and seem to be guided in a striking way by a providential power. They may, in various degrees, share the mind of God and be instruments of divinely empowered action in the world.

In the nature of the case, evidence that such divinely guided persons exist or have existed will largely depend on the testimony of others. It will always be strongly evaluative, since those who disbelieve in God or who dislike piety in all its forms will not assess the evidence in the same way. But for those who respond positively to their lives and teaching, these inspired persons will provide evidence of superhuman or supernatural influence. They may, to various degrees, in themselves be miracles, or 'objects of wonder', living astonishing lives beyond the normal powers of humanity and of great religious significance.

The word 'miracle' is, however, most often used of events, not people. So it is most often used of the extraordinary deeds and events that surround the lives of prophets, teachers and saints and that convey a sense of religious significance to others. Perhaps that is all we should expect of divine action in the human world, that it should transform the lives of those who turn to God, making them mediators of extraordinary goodness and wisdom, significance and power. That would be evidence enough for the worldly activity of ultimate mind and would strongly reinforce our own much lesser sense of divine presence and power.

4. THE 'LEAP OF FAITH'

If we ask the question 'What facts are necessary to reasonable religious belief?' the answer probably lies partly in personal experience of the divine and partly in reliable testimony to the influence of the divine in historical lives and events. Allowing for the common human tendency to embroider and elaborate on accounts of such events, to amplify them with astounding legends and cosmic mythologies, it would still be reasonable to pay careful attention to such testimony when it is very widespread and often given at great personal cost. A reasoned, if sometimes cautious, assent is all the more justifiable if we are disposed to think, on more general grounds, that the world is grounded in ultimate mind. Such considerations might lead us to make that passionate commitment of loyalty to the divine as conceived in a particular tradition, even in face of objective uncertainty, that is so characteristic of religious faith.

This might be called a 'leap of faith', in that it is a practical commitment beyond what the evidence would compel any reasonable person to believe. But it is not a leap into irrational absurdity, based on no evidence at all. It is a reasonable commitment based on a consideration of the sorts of experiential evidence that are appropriate and on the importance of coming to a decision on such a great matter, however inadequate our information. That is no more than we require and accept in most of the great personal decisions that mark our earthly lives.

5. BEYOND THE POWERS OF NATURE

Many religious traditions, however, claim a stronger sense of miracle than this. Miracles were defined by Pope Benedict XIV as events exceeding 'the power of visible and corporeal nature' (*De Servorum Dei Beatificatione et Beatorum Canonizatione, iv; De Miraculis*, 1783, 1.1.12). In this sense, miracles are not just extraordinary and beyond the normal. They completely transcend the natural powers of objects. They are, in a particularly provocative phrase of David Hume, 'violations of the laws of nature' (David Hume, *Enquiry Concerning Human Understanding*, Section 10, 'Of Miracles'; reprinted in an excellent collection edited by Richard Swinburne, *Miracles*, Macmillan, London, 1989, p. 27). If we dispense with the unfortunate connotations of the word 'violation', we can say that miracles are events that laws of nature cannot explain.

It is this stronger sense to which some scientists object. On 19 July 1984, the leading British science journal *Nature* published an article that stated that miracles are 'inexplicable and irreproducible phenomena [which] do not occur – a definition by exclusion of the concept' (p. 171). A number of professors of scientific subjects in British universities wrote to *Nature* to object to the arguments of the article. It is absurd, they rightly said, to define a miracle as something that does not occur and then say that, as a result, there are none.

Miracles are not totally inexplicable; they are just not explicable by known scientific laws. They are not irreproducible, but, since only God can reproduce them, they are beyond the powers of science to reproduce.

But are we to believe that an event not explicable by known scientific laws and not reproducible by scientists cannot occur? That would be a leap beyond the evidence a great deal larger than most religious leaps. If there are extraordinary acts of God, it is obvious that they will not be reproducible by scientists. And there is nothing in any natural science that proves no event can happen unless it obeys a general law of physics.

It is perfectly in accordance with science to see the laws of nature as primarily 'idealised models' of how things regularly happen when all external factors are excluded. Or, as John Polkinghorne puts it, our actual physical world is of a 'subtle and supple character whose laws are yet to be discovered and to which our presently known constituent laws are but asymptotic approximations' ('The Laws of Nature and the Laws of Physics', in *Quantum Cosmology and the Laws of Nature*, ed. Robert John Russell, Nancey Murphy and C.J. Isham, Vatican Observatory, Vatican City State, 1993, p. 442). This does not undermine the amazing predictive accuracy of physical theory. Our mathematical representations do mirror nature in important ways, but, as Bernard d'Espagnat states, they may not give complete access to the objective causal powers of nature, which belong to a 'veiled reality' (*Reality and the Physicist*, especially chapter 12).

Limits to our capacity to control the environment, inability to obtain totally precise measurements of initial conditions and the practical impossibility of coping with the number and speed of the required calculations ensure that we cannot establish that the universe is bound by a closed causal web that rules out divine action as one causal influence among others.

What the physical sciences provide is a high degree of predictability in controlled or closed causal systems, together with a hypothesis that such closed and isolated systems are deterministic, wholly defined by initial conditions and a set of physical laws. It is a much more speculative generalisation that the more uncontrolled and open physical systems in the natural world are likely to be deterministic too.

An equally plausible hypothesis is that, though physical laws completely govern the behaviour of physical particles and states in controlled or isolated systems, when those systems interact with other systems or with wider environments of types that cannot be exhaustively specified in advance, the operation of the known set of physical laws alone does not completely govern physical behaviour. There could be causal influences affecting physical behaviour that arise outside any particular specified system (for example, in far and unknown regions of the universe). Among those causal influences could be the intentional acts of agents, including God.

If God acts, God brings about a physical state in order to achieve some divine intention. It is scarcely credible that God's intentions should always coincide, by a sort of pre-established harmony, with what a set of natural laws alone entails. It is possible that at least sometimes God's acts should bring about non-physically determined states of nature, that God should exercise a causal influence on how things go, an influence that cannot be expressed by any purely physical law.

If God creates the universe for a purpose, reference to divine intention will be an essential part of the explanation of why the universe exists. But it will not be part of a scientific explanation and so it can be ignored by explanations in terms of physical laws. After all, a non-embodied agent cannot be seen to be acting, so one may always deny that any agent is involved who would be able to make states exist for the sake of the values they express or make possible. Such values can be interpreted simply as natural consequences of non-purposive physical laws, or as chance occurrences that may sometimes appear to violate the laws of nature, though they do not really do so. Even their value may be seen as a purely subjective evaluation of such states, made by human beings. So it seems that teleological explanation, in terms of God's action in the world, can always be denied. But it may be real nonetheless.

One of the insights of modern science is that God will not act in ways that destroy the structure of law and freedom that characterises the universe. God will act only to advance the divine purpose. But suppose a main part of the divine purpose is that creatures shall come to realise the source of their being in God and come to know and consciously relate to God. If knowledge of God ever occurs; if, as knowledge, it is partly caused by the presence of God; and if knowledge is a mental state that has, as its concomitant, some brain state; then it follows that God is a part cause of many specific physical (brain) states in the cosmos.

Such particular actions, however, will normally occur within a more general process of exercising a continuing influence on the emergence of the future, and so such actions, though strictly speaking beyond the purely natural powers of objects, will not be miraculous.

6. EXTRAORDINARY ACTS

If there is divine causal influence, it must integrate with the law-like nature of physical reality in a rational way, and not be an arbitrary interference with an otherwise smoothly running system. There will be a divine influence for good that preserves the relative autonomy of nature and its probabilistic laws, and the freedom of creatures to accept or reject the invitation to respond to the divine presence. As creatures come to sense possibilities of conscious relationship to the divine, there may also be specific divine acts that initiate such relationship, acts that may make the presence and character of God apparent. Perhaps a finite response of faith and love may open the world to the dynamic power of the divine.

Some of these acts may be quite extraordinary, and they are the ones that are generally termed 'miracles'. The degree of extraordinariness, the extent to which they transcend the normal powers of objects, will vary. But they will be law-transcending events, extraordinary events manifesting divine causality that modify the normal regularities of nature. What is important, from a scientific point of view, is that such events should not undermine the general autonomy and law-like structure of the cosmos and that they should have a clear rationale – they must have a vital and intelligible role in realising some important goals for the sake of which the cosmos exists.

One of those goals might be the disclosure of the mind-like or spiritual basis of nature, and an actualisation of the potentiality of the physical to become a conscious and free finite manifestation, or series of manifestations, of such a mind. It is possible that there could be such miracles, which would not 'interfere' in a closed physical process but would show in occasional and extraordinary ways what one ultimate divine purpose for the process was. The intelligible order of nature is not undermined by occasional modifications of its laws, especially when such modifications have the function of showing what the basis and purpose of the order of nature is.

The scientific understanding of a universe of intelligible law and emergent creativity changes the perspective within which we see divine action. If we are to speak of divine action in such a universe, it will be in the context of a cosmic process of emergent value, within which humans can begin consciously to realise the spiritual potential of the physical. Some aspects of such realisation may have a character that we call 'miraculous': extraordinary and astonishing acts, beyond the ordinary powers of nature, that disclose the spiritual nature and goal of the whole cosmic process. That is the most adequate definition of a miracle of which I am aware. The scientific vision of the nature of reality provides a new context for the occurrence of miracles, but it does not render obsolete the idea of particular or miraculous divine acts.

7. EVIDENCE FOR MIRACLES

The eighteenth-century Scottish philosopher David Hume suggested that scientific evidence for the existence of laws of nature will always outweigh any claim that a miracle (an event not explicable by reference to such laws) has occurred, so that science prohibits rational belief in miracles (*Enquiry concerning Human Understanding*, ed. A. Selby-Bigge, Oxford University Press, Oxford, 1902 [1748], p. 114). However wonderful Hume is as a philosopher, this statement is quite unacceptable. Even an enormous amount of testimony that events have always obeyed regular laws of succession in the past cannot count against a well-grounded claim that there has been an exception to such regularity. Observation can confound our greatest expectations, and one good case of reliable observation provides just the counter-example that is needed to show that nature does not always obey what we regard as regular laws of nature. The whole point of the new scientific stress on observation, and the need to test hypotheses by looking for counter-examples, establishes as a scientific principle that an observed counter-example to the universality of laws of nature is something no scientist should be too surprised to discover.

In any case, there is not, as Hume suggests, universal testimony that laws of nature exist. Until scientists like Newton formulated them, people were not even sure that there were such laws. And Newton certainly did not think they were universal in scope and unbreakable in principle. Whether or not we think miracles occur is a matter of how we assess the evidence for them. Such assessment depends upon many factors. One is the reliability and intellectual maturity of the observers of miracles. David Hume argues that we should never rely on testimony to miracles, because 'no testimony is sufficient to establish a miracle, unless the testimony be of such a kind, that its falsehood would be more miraculous, than the fact, which it endeavours to establish' (*Enquiry*, section 10). The probability that the testimony to the occurrence of an event is false must be less than the probability that the event occurred. But, Hume says, the probability that miracles occur is zero, so no probability can be less than that and consequently no testimony to such an event can ever be accepted.

This is not the case. The probability of miracles is not zero. There is always some possibility, if there is a God. It is very difficult to assess their probability, since that probability depends upon the background facts. If there is a God and it is important that finite beings should know what God's purpose for them is and only an extraordinary event can assure them of that, the probability of a miracle will be quite high. Since we cannot be sure of any of these things, we can only say the probability is unknown. But it is not zero and it is not 1 (as high as possible).

The probability of mistaken testimony also varies widely. The testimony of one's own eyes, backed by others in good conditions of observation, is very high, though never absolute, since we may be tricked by a good magician. Most testimony to miracles could well be mistaken, though we might well think that some is reliable, where the witnesses seem honest, well balanced and morally estimable.

So what we typically have is an extraordinary event of unknown probability balanced against testimony that could be mistaken but is unlikely to be always mistaken. In this case, the probability of being mistaken about the occurrence of a miracle could well be less than the probability that the event actually occurred. If the reported miracle is the sort of thing a specific believer in God might well expect, and the witnesses are trustworthy and reliable, it would be reasonable to believe in miracles, on Hume's own principle.

But it is worse than that for David Hume. We could easily have a case in which the probability of deceit or falsehood is greater than the improbability of a miracle or extraordinary event and yet we would be rational to accept the testimony. Suppose I win the jackpot in the National Lottery. The probability of that happening is extremely low – in 2005 it was 1 in 13,983,816. Suppose a friend goes to tell others about my win. The probability that he is lying or has misheard me may be low, but it is not zero. We may put it at around 1 in 20, say. So the probability my friend is mistaken is much higher than the probability that I won the lottery – which contradicts Hume's principle. Yet it would be reasonable for his wife to accept my friend's testimony to this hugely improbable event. Hume's principle is unacceptable. It might even be unreasonable for a believer in God to discount all testimonies to the occurrence of miracles, where they are widely reported by the wisest and most committed believers.

We should scrutinise reports of miracles carefully, because people are known to exaggerate, be deluded or be victims of deceit where miraculous events are concerned. We should not, however, discount seemingly strong testimony just because, much less to the extent that, the events reported are very unusual. A report of a maximally unusual event can be rationally accepted if the report meets the tests for delusion and deceit that we usually apply. And miracles, for a theist, are not even maximally unusual. They are irregular, rare and exceptional. But good reasons might be found for their occurrence, insofar as they fit into a general web of beliefs that leads one to expect physical-law-transcending actions, unusual or not.

8. TYPES OF DIVINE ACTION

Three main forms of divine action have been distinguished. First, there is the general action of creating and holding in being a universe. This action selects the basic parameters and constants of physical law.

Second, there are actions shaped by the controlling intention to encourage a community of sentient and moral agents to come into existence. These acts will be within the probabilistic limits of physical law, and will normally be influencing, but not sufficient, causes of processes tending to the emergence of values. The physical sciences can ignore such divine actions, since it is possible, even if often improbable, that they happen simply by chance.

Third, there are specific acts in which the normal physical powers of objects are transcended. God is always acting to sustain the universe in general and to guide the emergence of new forms of value in co-operation or sometimes in conflict with the acts of creatures. But God may also act occasionally in miraculous ways, leaving the probabilistic processes of nature and the autonomy of creaturely choice intact, but also manifesting in an extraordinary way in the physical realm the underlying spiritual basis and ultimate purpose of the cosmos.

9. DECISIONS ABOUT MIRACLES

My own general view is that not too much importance should be placed on the occurrence of miracles, in the sense of divinely caused and utterly astonishing events that are totally inexplicable in terms of physical laws. The reason is that there are many recorded cases of deceit and delusion, many examples of human gullibility and wishful thinking and many instances of hyperbole and exaggeration in accounts of the lives of saints. So it is wise to be cautious in accepting testimony to the occurrence of miracles.

It must also be allowed that if present experience is to be a measure of the probability of the occurrence of far past or far away events, more modest claims of divine guidance, inspiration or healing seem to be more likely to be accurate than claims of amazing events a parallel to which we have never ourselves seen.

Nevertheless, since miracles are rare by definition, we may feel there is convincing testimony to events of a sort we are unlikely ever to experience, but which confirm decisively important truths about the nature of ultimate mind and the ultimate goals of the cosmos. A properly scientific attitude should be open to that possibility, while perhaps refraining from coming to a decision, on strictly scientific grounds, as to the occurrence of reported miracles.

How, then, might a decision be made about whether a miracle has occurred? At this point we return to the question raised in chapter 1 about the relation between faith and fact. There it was accepted that where scientific study can establish the facts (in the case of the age of the universe, for example), religious faith must simply accept the verdict of science, however provisional it may be. But in the case of historical claims, especially about alleged divine causality or miraculous events, the facts cannot be established in a neutral way. A whole interconnected web of beliefs comes into play – beliefs about whether God exists, what the character of God is, what sort of acts such a God might perform, what the character of a religious tradition that reports such facts is, how far we approve or disapprove of that character, and what sorts of personal experience of God we might have. We are relying on personal experience, shaped within a particular culture and tradition, on the reported experiences of people we like, admire or trust, on general beliefs about the nature of the world, and on our evaluations of the institutions and ways of life we have encountered.

This is all what I have called experiential evidence, not observational evidence, so it has an inescapably personal character. To make a rational decision in such an area requires one to gain as extensive an experience as possible, to make one's evaluations as consistent and unprejudiced as possible and to be sensitive to as many diverse viewpoints as possible.

At this point the objectivity of science gives way, quite properly and inevitably, to personal experience and evaluation. Science as such does not enter this area, though it provides the necessary factual background for it. That is perhaps why many scientists are sympathetic to the idea of ultimate mind, but cannot see any specific religious implications of the idea. For religion usually calls for a commitment to some specific tradition of reported divine revelation, and for loyalty to a person or persons who are claimed to have special insight into the divine mind and to have become in a special way instruments of that mind in the world. Yet, once the existence of ultimate mind has been accepted, to look for the appropriate sort of evidence for its intentional actions in the cosmos becomes an integral part of that search for knowledge which drives the scientific enterprise. In that sense acceptance of and reflection on the scientific worldview lead naturally and reasonably to the pursuit of religious truth.

16

IMMORTAL LIFE

*The ultimate purpose of the universe is not likely to lie at its phys-
ical end. Though some cosmologists foresee the possibility of an
endless physical life in the cosmos, most see all life as coming to an
end. Yet scientific speculations about the far future universe open
up the possibility of a continuation of life in different forms and in
different space-times. Although the purpose of the universe may
lie in its mid-life maturity, not its end, there may yet be a possibility of a fulfilment of life beyond this physical cosmos.*

*AIM: To show that modern science opens up speculative
possibilities of conceiving immortality in different realms of being
and forms of embodiment. Materialism seemed to close off the
possibility of immortality, but modern science opens it again.*

1. INTO THE FAR FUTURE

Much modern scientific speculation is concerned with the
nature of time, the status of the fundamental laws of physics, and
the way in which purposive actions mesh with the apparently

non-purposive causal processes of the material world. These are all perplexing problems on which there is wide dispute and no agreed resolution. The same problems arise with regard to belief in God, and the same disputes exist about whether God is timeless or temporal, whether God acts outside the laws of physics and how God's acts are to be understood in a world of physical laws.

The situation is not that all scientists are agreed on one set of answers to these problems and have settled on a materialist view that excludes God altogether, whereas believers have ranged themselves against science and settled on a position that ignores science completely. The problems are common to contemporary science and contemporary religion, and the spread of opinion is very similar in scientific and religious circles.

I have argued for a temporal view of God, an 'idealised model' view of laws of physics, and an account of divine action as transcending but not contradicting the laws of physics, properly understood. I have appealed to some of the insights of modern science in doing this, but I am not suggesting that all scientists, or all believers either, will agree with what I have said. I hope it would be agreed, though, that there is much in modern science that supports the view I have taken, and that this view is well enough established to provide a reasonable and coherent account of how the God of religion can accord with a modern scientific account of the universe.

The final decision about whether or not to embrace a commitment to one path of seeking personal knowledge of, or union with, God will no doubt be based on personal experience (and perhaps, believers would say, on the gift of faith). But for many of us it is important to know that belief in God is a rational option, which completes the scientific quest for understanding the universe and does not compete with it.

One major issue remains to be discussed. I have written about the origin of the cosmos and about the long evolutionary process that I have called 'cosmic ontogenesis'. I have written about the possible active part that God might play in present interaction with human persons on this planet. But both scientists and religious believers have an interest in the question of the end of the universe, and of its ultimate goal, if there is one. What sort of interaction exists between science and belief in God when one considers these matters?

2. THE END OF THE UNIVERSE

It is certain that life on earth will end. In about five thousand million years, astronomers presently estimate, our sun will die and the earth will die with it. Yet there will be a vast amount of time still to go in the universe after all life on earth has ceased to exist. Perhaps our descendents will have learned how to leave the earth and live on other planets around other suns, or in the space between the stars. How long will such life continue? Is it too bound to end eventually?

In 1854 Hermann von Helmholtz predicted what has been called the 'heat death' of the universe. In accordance with the second law of thermodynamics, any closed thermodynamic system will progressively increase in entropy (disorder) until at last there will exist nothing but space without any order or organisation at all. All complexity, including all life, will inevitably cease to exist. It looked as though life was inevitably going to end.

This picture has since been revised, however, to provide a number of possibilities of 'endless' life. They may seem rather fantastic, or even undesirable, but it could be that a sort of physical immortality is not impossible. Some speculative mathematical physicists have supposed that intelligent life may one day take over the whole physical universe, and even 'spread into all spatial regions in all universes which could logically exist, and will have stored an infinite amount of information, including all bits of knowledge which it is logically possible to know' (Barrow and Tipler, *The Anthropic Cosmological Principle*, p. 677). That would indeed be the transformation of the physical into the mental. Could it be the ultimate goal of this, and of every possible, cosmos? There are reasons to think it could not, for if the universe is accelerating, as most cosmologists now think (this week, anyway), huge parts of it will never be 'mentalised'. So at best the possibility of a sort of endless life could only exist in some (collapsing) regions of the universe.

What will happen in the far future of the universe is far from certain. There is an immense amount that we do not know about the universe, and there may be forces beyond the present horizon of knowledge that will become important in time and of which we can now know nothing. We do not know if the fundamental constants of nature may change. We do not know whether this space-time may be just a tiny bubble in a wider multiverse whose laws and powers are quite different from those we know. Nevertheless there is much fascinating physics that has recently become relevant to the question of the end of life and of the universe. Freeman Dyson put the topic on to the physics agenda with a paper entitled 'Time without End', in *Reviews of Modern Physics*, in July 1979. This paper is reprinted in *The Far-Future Universe* (ed. George Ellis, Templeton Press, Philadelphia, 2002), along with a paper ('Life in the Universe') originally written in 1981 and revised for publication in 2002 which re-affirms his central case, in response to published criticisms.

Dyson distinguished four possible sorts of universe – closed, decelerating, open and accelerating. In the closed and accelerating universes, he thinks that survival is impossible. But in the decelerating and open universes, he thinks a sort of endless life is possible (*The Far-Future Universe*, p. 156).

(i) A closed universe

If the universe is closed, with finite volume and duration, it will expand to a maximum size and then begin to collapse again. It will end with a 'big crunch', in which all possible forms of life will be totally fried by rising temperatures just before the end.

This sounds depressing. However, Barrow and Tipler have argued (*The Anthropic Cosmological Principle*, pp. 660–77) that, as the universe contracts, time will (subjectively) go faster and faster. Any living beings will be able to have more and more thoughts in a shorter time. As this process is exponential, in the last few seconds before the big crunch, any living beings that then

exist might be able to have an infinite number of new thoughts –
'an infinite number of physically distinct events happen on
approach to the singularity' (p. 606). It could be that an information-processing entity 'would process an infinite amount of
information' in an objectively finite time. This would seem like
endless life. So even in a closed or recollapsing and dying universe
there could be a sort of endless life. Such life would be very different from anything human, but it would be life of a sort. The
human species, however, seems certainly doomed to extinction.
In any case, the hypothesis of such a big crunch 'seems to be
inconsistent with recent measurements' (Dyson, 'Life in the
Universe', p. 147). So the chances of physical immortality seem
rather slim so far.

(ii) A decelerating universe

If on the other hand the universe is not closed but continues to
expand, yet decelerates as time goes on, the outlook is better. As
the velocity of distant galaxies decreases, if there are unlimited
reserves of gravitational free energy, humankind should be able
to exchange material resources and information with the most
distant galaxies. It might take immense amounts of time, but we
would have the time. The universe will expand more and more
slowly until it simply stops. At that stage it will have run out of
energy. But time will then be going so slowly that in subjective
consciousness things will seem to go on for ever. Again, subjective endless life is compatible with an objective 'end of time', as the
universe slowly enters into the big freeze and grinds to a halt.

(iii) An open universe

Dyson defines an 'open universe' as one in which there is no
deceleration, but expansion continues linearly (at a more or less
constant rate) for ever. In such a universe, he argues, life would
not have access to an infinite amount of energy, since distant
galaxies would continue to recede and we would not be able to
exchange matter with them. Nevertheless life could appear to

survive for ever using a finite amount of free energy, as long as life adapted to the changing conditions by lowering its temperature and conserving energy by such devices as long periods of hibernation.

In all these contexts, life has to be defined in terms of information processing. A conscious living being is one that processes information. If some form of life can process an infinite amount of information in a finite time, it will subjectively seem to live for ever. In the expanding universe scenario, in thousands of billions of years all that will be left is a soup of photons, neutrinos and gravitons moving slowly in a slowly expanding space. Normal matter will have ceased to exist. Information storage and processing could still continue, however, using lower-energy quanta of energy, or radiation at longer and longer wavelengths. In other words, an infinite amount of information could be processed with a finite expenditure of energy. Given the material conditions of the far future universe (or lack of them), thoughts and memories would have to be downloaded into magnetic fields of clouds of photons and gravitons – but we have thousands of billions of years to learn how to do that. Living beings (information-storing and-processing devices) would think very slowly, but they could think for what seemed to them like endless time, in between their increasingly long periods of hibernation.

(iv) An accelerating universe

The final scenario Dyson considers is that of the exponentially accelerating universe. Here galaxies move apart with ever increasing speed and humanity would eventually become marooned on a isolated island in a universe with which no further exchange of energy or communication would be possible. He sees no hope of even a subjectively endless life in such a universe – which is, regrettably, the one most favoured by cosmologists at present. John Barrow has also become much less optimistic about the possibility of endless life in an accelerating universe. 'Along our world line, in our part of the universe, there will ultimately be sameness, starless and lifeless forever, it seems' (John Barrow,

'The Far, Far Future', in *The Far-Future Universe*, p. 38). In what seems like a move of cosmological desperation, an extreme strategy to avoid this outcome is proposed by Lee Smolin (*The Life of the Cosmos*, Oxford University Press, Oxford, 1997) – that the collapse of stars into black holes may create new universes. Events within a black hole are isolated from contact with the rest of the universe, owing to the immense gravitational force within the hole. But new forms of space and time may form within such holes, and so our universe may generate many other mini-universes within itself. With a future technology far more advanced than our own, it might be possible to predetermine the structure of such black hole space-times. It might be possible for our remote descendants to decant themselves through a wormhole into a new baby universe they had themselves designed. They could continue doing this for ever, moving from one universe to another without end. Indeed, who is to say that they may not already have done so and that they are not already here? The immortals may move among us, in the darkness between the stars, and in time we may hope to join them, when our evolutionary journey in this cosmos is ended.

3. IMMORTALITY WITHOUT WORMHOLES

Physicists, not fantasy writers, are propagating these far future scenarios. They are highly speculative hypotheses, but they are founded on well-understood and well-established theories in modern physics. What these hypotheses suggest is not that these things will happen, but that they are possibilities allowed by modern physics. This cosmos may die, but it is just possible, at least in some forms of far future universe, that intelligent life may exist for ever, or at least may subjectively seem to exist for ever. Is that of any real importance to us? Charles Darwin, writing in 1859, said that 'we may safely infer that not one living species will transmit its unaltered likeness to a distant futurity' (*Origin of Species*, p. 489). Humanity is not necessarily the 'crown of creation'. Indeed, it is very unlikely to be the highest form of conscious life there will ever be. We will probably become extinct as a species, superseded by more intelligent and durable life-forms.

Information-processing clouds of particles in intergalactic space and silicon-based computers into which thoughts and memories have been downloaded (another possibility of escape from a dying sun) are not human. Even if some form of intelligent life will survive for ever, it will be more different from human life than we are from dinosaurs. But I suppose the thought is that at least the universe will have a goal, a final flourishing of intelligent life that will make it all worthwhile. Two sceptical thoughts immediately arise. First, is it more than the most tenuous hope that such a goal will be achieved? Second, is the attainment of such a goal really necessary, and is it in any case really sufficient, to make the existence of the universe worthwhile?

Martin Rees, in *Our Cosmic Habitat* (Princeton University Press, Princeton, 2001), evinces moments of long-term cosmic optimism. 'Our remote descendants are likely to have an eternal future' (p. 121), he says, though without much conviction. Even this cautious optimism is rather cancelled out by his belief that humans will probably extinguish all life on earth long before these cosmic scenarios become relevant. 'Extreme pessimism seems to me the only rational stance,' he says ('Living in a Multiverse', in *The Far-Future Universe*, p. 83). The prospects of human survival even into the next millennium can seem very bleak. The population explosion, the existence of nuclear and biological weapons that can destroy all human life on the planet, the depletion of energy resources, and the growing gap between rich and poor parts of the world all pose new and severe threats to human existence.

Whatever we think about the far future of intelligent life, we will each of us die very soon, and most of us humans will have had little or no opportunity to live in a way that enables us to use our capacities creatively, to enjoy the many possibilities of goodness and beauty the earth affords and to live in harmonious friendship with our fellows. Intelligence may outlive the stars, but I will die soon, my hopes unrealised, my plans abandoned and my life diminished by suffering and pain. It looks as though this planet is a failed experiment, soon to be abandoned. We will die, unfulfilled and unlamented, and true intelligence, if it is to exist at all, will grow and develop elsewhere in the cosmos. Would that really constitute the achievement of a goal that would make the whole universe, including our lives, worthwhile?

Before total depression takes hold, we should pause and take note that most of these cosmological theories, however exotic, are based on the premise that intelligent life must have some physical basis in this universe. When that physical basis has been eviscerated so that all we have left are information-processing mechanisms composed of sub-atomic particles in empty space, we may well wonder whether the quality of experience – of feeling, evaluation and awareness – possessed by such beings has been diminished below a level that would be of any interest to us.

The best hope may be that of decanting ourselves into another universe, where the conditions of life may be better and could even be designed by us to order. In that way, we might be able to escape from a slowly freezing and almost empty universe, by disappearing down newly constructed wormholes into a much more comfortable designer universe.

When we have got used to such bold speculations, it is not a much greater step to move on to the hypothesis that a consciousness that originates in one universe might be transferred to another universe instantaneously, without the tedious necessity of passing through a wormhole. Those who believe in God believe that there exists an unembodied primordial mind that could be the origin of many universes and might well have the power to initiate such an instantaneous transfer of finite persons between them. By the power of God, we might pass into an alternative universe without the inconvenience of having to find a nearby wormhole. Whether or not God could or would do such a thing cannot be established by the natural sciences, which on the whole mercifully refrain from saying much about what God might do. But at least contemporary physics seems to have opened up in a rather unexpected way the bare possibility of a continuing life beyond this universe.

4. A SPIRITUAL ALTERNATIVE

The interest of speculation about the end of the universe, apart from the fascination of the physics involved, is an interest in the ultimate future of living beings. Is there a worthwhile future for living beings, an ultimate goal that is supremely desirable and attainable? I doubt if the existence of almost unimaginable information-processing units in some far future quite meets those requirements. There would have to be a quality of experience that is appreciated and enjoyed for its own sake. It is rather doubtful whether we could think of intergalactic clouds in an increasingly cold and inhospitable space as having a sufficiently rich physical basis for enabling them to create, appreciate and enjoy beauty and friendship. It is possible, but it seems more likely that the intrinsically desirable life, if it exists, will exist not at the end of this physical process, but at its mid-point. That is what the model of cosmic ontogenesis might lead us to expect – a cosmic organism that grows to maturity and then decays and dies.

The history of the cosmos, as I have presented it, is an expression of ultimate mind, bringing into being by a long evolutionary process other minds, physically based and physically generated, that can share, in an appropriate way, the self-awareness and creativity of their originator. In this sense, the cosmos does move towards worthwhile states of consciousness. But within the cosmos such worthwhile states may exist maximally at the mid-point of the life of the cosmos, in its mid-life maturity, and not at its end. Why should the life of the cosmos not be like human lives, developing towards a state at which many potentialities of experience could be actualised, and then slowly decaying into quiet oblivion? The goal of the physical universe is perhaps not that it should end with a grand climax of intelligent consciousness, but that it should at appropriate points and in appropriate ways generate from itself personal beings that can manifest the experience of the one eternal God in new ways and enable God's creative action to be expressed in new ways. The goal, in other words, would be not at the end, but in the process as a whole.

Yet there is an attraction about the idea that the things we have done and experienced should not be lost, that somehow there might be a future in which all – or at least all the good – that we have experienced might be retained. And we might naturally wish that we humans could participate in the mature and worthwhile future of the universe, which looks as though it is otherwise going to be reserved for some far future species not very like us.

If there were an ultimate mind, perfect remembrance would be assured. For in a mind possessed of the greatest possible knowledge, nothing that has been will be lost. Whatever happens to the cosmos at its end, the good things that have existed during the course of its existence will be held in the divine mind for ever. This is what A.N. Whitehead, in his process philosophy, calls 'objective immortality' (*Process and Reality*, p. 347). For many people it is enough that what they have done and experienced will be remembered for ever.

Yet on this view *we* would not share in the divine memory of all that has been, and of ourselves. Moreover, much in our lives will have been left unresolved. Many possibilities will remain unrealised, many paths unexplored and many experiences will have been disfigured by frustration and suffering.

If our cosmic existence is a journey towards a realisation of our own distinctive capacities and a full appreciation and integration of our own unique life history and experiences, this journey is, for most of us, radically uncompleted. What we have done and experienced is important but incomplete. Might there be a possibility of some completion by relation with an ultimate mind that contains the record of all human experience and the capacity continually to call new futures into being?

What far future cosmological speculations suggest is that it may be possible for living beings, with thoughts and memories, to exist in very different forms than the ones we are familiar with, whether in intergalactic clouds or in silicon artefacts. We ourselves could exist in different physical forms, our thoughts and memories downloaded into some other piece of hardware. We may even exist in different space-times, in realms where the laws of physics are quite different. If we can envisage travelling through a wormhole into a different space-time, it is not much more difficult to envisage simply being transferred by the greatest super intelligence there is, by God, to such a space-time without the bother of going through the wormhole. The idea of a life 'after death', in some other sort of universe, when these physical bodies have decayed, comes to seem much more plausible in the light of these far future speculations of physics.

5. THE POSSIBILITY OF IMMORTALITY

Where there exists the sense of a continuing self, enduring through a series of acts and experiences, the possibility arises of a continuance of such enduring subjects beyond the decay and dissolution of their particular bodily forms. Even in the case of DNA molecules, the genetic code can be separated from its material embodiment. The code will usually be embodied in some material form, but it is not confined to the form in which it originated. It can be embodied in a totally different material way, while remaining the same genetic code that arose through evolutionary processes in nucleic acids.

Insofar as minds largely consist of information content, that content could, in a similar way, be transferred to a different material embodiment. One could have a continuing sense of a self that has been shaped by past experiences, that has many memories of the past and plans for the future, that accepts responsibility for its actions and resolves to seek specific goals in future, but that has been transferred to a different body.

In one sense, this happens naturally anyway. The cells of the body constantly die and are replaced. Our bodies change gradually over time and are very different when we are old from how they were when young. So we all inhabit changed bodies in the course of our lives, though we normally expect the changes to be gradual and continuous. Sometimes, however, catastrophic changes occur to some people, through accident or disease. We can already have artificial limbs, and it is possible in principle to have artificial internal organs. So some day persons may be able to have completely different, artificially constructed, bodies. These improved bodies would still, however, have a physical continuity with their replaced bodies.

Is such physical continuity essential? Modern scientific knowledge suggests that it is not. It is theoretically possible to duplicate the exact physical configuration of a human brain and body some time after the death and decay of the original. This duplicate could be indistinguishable from the original. If it remembered the same things, had the same character and carried on with the same plans, we would probably say that the same person had been reconstituted, or perhaps 'resurrected'.

Frank Tipler (*The Physics of Immortality*, Doubleday, New York, 1995) has argued that the vast intelligences that might exist in the far future universe, possibly as clouds of dust in intergalactic space, would be able to 'resurrect', or create copies of, all living beings that had ever existed in the whole history of the universe. There could exist exact replicas of each of us, and it would seem to us as though we had been brought back to life.

There would be a complication if many duplicates were made, since they could not all be the same person. We would have the problem that Captain Kirk had in one episode of the television series *Star Trek*, when two of him emerged from the matter transporter after only one of him had gone in. Which would be the real Captain Kirk? In that episode, as I recall, the problem was solved because one of them died – and of course that one was not the 'real' Captain Kirk. But the simplest course would be just to say that there would now be two people who used to be Captain Kirk. Captain Kirk, like an amoeba, has split in two.

There could in fact be many genetically identical individuals who used to be one person, but have now split into many (identical twins are different persons who have split from the same original fertilised cell). That would not be a great problem, though it might be difficult to say which one of them owned the family car or was married to a probably rather bewildered wife.

So it seems that we, the very same selves we are, with the same memories, feelings, hopes and fears, could exist again after the death of our bodies. A greatly advanced scientific technology could enable us to exist again in bodies with no physical continuity with and very unlike our present ones (perhaps constructed out of more durable materials like silicon). Exactly how unlike our present bodies a future embodiment could be is at present unclear. Presumably, our future bodies would have to be able to provide information we could assimilate with our present knowledge, and provide possibilities for action that we could relate intelligibly to our present goals and desires. But we could probably cope with major changes in environment and somatic constitution and might well be glad of any physical improvements that scientists had devised.

Tipler's hypothesis that there might be future replicas of ourselves may have some attraction. Yet that hypothesis has drawbacks, for if we were simply replicated, we would just carry on with the same sort of lives we have now, with all our forgetfulness, folly and failure to be what we could be. What is required is the recovery, the perfect remembrance, of all that has been, together with some form of continuing life that would yet be transformed so that we would be freed from the petty hatreds and egoisms that would make us incapable of fitting into any fully mature and worthwhile society. That requires something more than replication, and it almost certainly requires something more than a future re-embodiment in this physical cosmos. It requires something that will take us beyond the realms of physical science altogether.

17

THE ULTIMATE GOAL

The God of religion is a reality of supreme value that can be known by finite minds. That knowledge can transform and fulfil their lives. The religious vision of the cosmos transcends science by speaking of an ultimate mind that enters into the experiences of finite beings and unites them in knowledge and action to the divine mind. This points to a fulfilment of life beyond the cosmos and to the defining function of religion as leading to union with ultimate mind.

AIM: To show that belief in God implies the possibility of endless life with God beyond this cosmos, but also the vital importance of this cosmos to the genesis of the embodied selves that human beings are. To show that religion is, at its best and most authentic, a discipline of the mind that aims at the fulfilment of personal life as it is transformed by the presence and power of ultimate goodness.

1. BEYOND THE COSMOS

In a previous section (8.1) I pointed out that we could only have come to exist in this entangled cosmos and that we could not have existed as the precise individuals we are in another cosmos. Our inner lives provide us with perspectives on just this cosmos, of which we are essentially part, and our actions are causally connected to the possible paths into the future that this cosmos provides.

Memories, feelings and thoughts are all embodied in physical mechanisms of the brain (rather as Beethoven's Ninth Symphony is embodied in a binary code on a compact disc). I have argued that such feelings and thoughts are in principle separable from the brain and can logically exist without such embodiment (as Beethoven's Ninth can exist apart from the compact disc). But if they did, life would be unintelligible. We would not be able to tell how or why such experiences originated, and all causal connections with a real physical environment would be disconnected. So it is a good thing that experiences are embodied in a real and objective neurological, physical and social context.

The point is, however, that they could be so disconnected. There could be the sense of a continuing self with memories, feelings and thoughts, though now there would be no world with which the memories could connect or in which actions could be performed. Such a solipsistic dream world, however, would not allow for any real development or continuation of a fully personal life as an agent in interaction with others and with an environment that could provide new knowledge and opportunities for action.

An after-death dream world is possible, but it is not the continuation of a fully personal life. An attempt to construct a sort of social dream world is made by the philosopher Henry Price in *Essays in the Philosophy of Religion* (Clarendon Press, Oxford, 1972, pp. 105–17). It is worth noting, however, that in Price's dream world, persons have to imagine themselves with bodies, even if their bodies are rather dreamy. It becomes difficult to distinguish dream and reality if many people can inhabit the same dream. In fact, the dream seems to turn into a sort of embodied world, though of a very flexible and unpredictable sort.

Some religious traditions – notably the Tibetan *Book of the Dead* – do speak of such a dream world as an immediate stage after death. In the dream world the self can work through its unfinished psychological business, its unresolved problems and some of the consequences of its earthly actions. However, this would be more like coming to terms with what had been done and experienced during personal life than a continuation of personal life. In that sense, death is a real ending to the sort of embodied existence human persons properly have. It is a radical end and not merely a change of location.

Nevertheless, having become the persons we are in this cosmos, could we not be replicated in a different space-time, free of the entanglements of this space-time, to begin a new fully personal life there? If we can be replicated in different forms in the far future of this universe, in galaxies or intergalactic spaces far away, it is not really a much bigger leap to be transferred into another universe.

These may seem extreme hypotheses of purely speculative physics. But they are on the map of possibilities, and modern physics has opened up pathways of imagination that a more pedestrian materialism thought science had debarred. If we can imagine future travels into another universe, then we can certainly conceive of our replication in another universe, another form of embodied existence unlike this, and beyond all possibility of communication with or return to this cosmos.

It is conceivable that we, born in and shaped by this sublunary realm of suffering and decay, might come to exist in universes beyond this one, perhaps endless in number and variety. There we might find our sufferings and failures transformed, and our hatreds and self-centredness burned away, as we come to realise the harms we have caused and the futility of the paths we have chosen. Our frustrated potentialities could perhaps be realised in new forms, and our understanding and creativity expanded in an endless journey into a fullness of being beyond what we now know as human personality. This is very much the vision sketched out by the fourth-century Christian writer Gregory of Nyssa, but it can be shared in some form by most monotheistic faiths.

I do not think science can pronounce on this possibility. The extraordinary thing is that modern science suggests ways in which such a possibility could be consonant with, and even a plausible extension of, our scientific understanding of the universe.

If this universe is grounded on a reality of supreme perfection, and if the universe involves the sorts of suffering and frustration of purpose and potentiality that it does, then it seems plausible that a perfect origin of all things would create some form of existence in which earthly sufferings and potentialities could be transformed and fulfilled. If the ultimate cosmic mind has complete knowledge of all that happens in the universe, and if that knowledge remains in the divine mind for ever, then it seems possible that such knowledge could be shared by finite persons, either now or in a further existence. That is the ground for the hope that there might be a form of immortality in which finite persons could share more fully in the life of the divine mind, and in which ultimate mind itself could find a novel creative expression of its nature by entering into unending relationship with a society, and perhaps with indefinitely many societies, of minds.

2. THE VALUE OF PERSONHOOD

The hope for immortality is not, as Karl Marx supposed, just a hope for a better life after death which makes the sufferings and activities of this life relatively unimportant: the 'opium of the people'. Immortality is important precisely because personhood in this world is of eternal significance. From the point of view of someone who believes in a creator God, to be a person is to be a self-conscious agent who can grow towards the ultimate good by creative striving, by the use of responsible freedom, and by the development of both a unique and irreplaceable form of experience and a self-shaped realisation of unique inner potentialities. In the perspective of eternity, each person adds a distinctive contribution to the divine experience, to the content of ultimate mind. Each person mediates divine action, the active expression of ultimate mind, in an individual and creative way. And each person could well be destined to an unending relationship with a spiritual reality that values that person for their own sake and is able to give them a share in the divine life for ever.

The religious hope for immortality is based on the belief that individual personhood in this cosmos has such enormous significance and value that a wise and powerful creator of the cosmos would ensure that persons will find in a life beyond this cosmos the fulfilment and relationship that is so often frustrated in this cosmos. The value and dignity of human (and of all forms of) personhood, and the importance of what persons do and experience, are not undermined, but are affirmed and strengthened by the hope for immortality.

Belief in a life beyond death is based, as far as speculative thought goes, on an affirmation of the eternal significance of personhood and on belief in a perfect self-existent spiritual source of all being. The natural sciences do not show that souls are immortal. Insofar as the sciences deal only with the physical, and with the publicly observable elements of personhood, they may even be taken to throw doubt on the possibility of immortality, which entails a disconnection of the self from a particular physical body. I have suggested, however, that recent scientific interest in information theory, in the possibility of a multiverse and in the admittedly remote technical possibility of duplicating mental content in other forms opens up the possibility of personal life after death in a new way. If there were some corroborating evidence for this possibility from other sources (it would have to be what I have called 'experiential evidence', probably based on alleged revelations or extraordinary disclosures of the divine purpose to a spiritually oriented person or group), the sciences could certainly not discount it.

3. TOWARDS A CONCLUSION: THE LIFE OF THE COSMOS

It is sometimes thought that the natural sciences presuppose or support a materialist view of the cosmos. For such a view, speculations about spiritual reality are idle, and any thought of humans existing as anything other than organic physical bodies is fantasy.

But there is another view. As a matter of historical record, most of the great classical scientists have held a spiritual view of the world, a belief that the whole material world is the product of creative intelligence. In the past century, modern physics has cast doubt on the very conception of 'matter' as independently existing stuff. In quantum mechanics, in particular, physics is seen to be dealing with a 'veiled reality' whose inner structure can only be grasped mathematically, and with a cosmos in which matter is only one of many forms energy can take. The cosmos is seen as having a history in which intelligence emerges from unconscious interactions of particles and waves. Intelligence may, in the form of information storage and processing, be seen as a final intrinsically worthwhile goal of the whole cosmic process. Modern physics makes possible a new vision of the cosmos as an intelligible, rational and mathematically beautiful whole, developing through processes of creative self-organisation towards new levels of complex integration that enable it, or beings within it, to become self-aware and self-directive.

It is plausible to see this process as planned by a supreme intelligence that put in place the cosmos's intelligible structure and directs its progress towards the final goal of conscious knowledge and co-operative action. The knowledge and power of beings within the cosmos, or even of the cosmos itself if it ultimately becomes, as some speculatively claim, the physical basis of one or more super-intelligent information-processing organisms, may become almost unimaginably extensive. But such knowledge and power will still be the product of the mind that put the process in place. In that case, the story of the cosmos will be the story of a mind that actualises the cosmos for the sake of its beauty and also for the sake of actualising within the cosmos finite minds that can experience and creatively add to that beauty.

4. THE RELIGIOUS DIMENSION

This story takes on a religious dimension insofar as finite minds are thought to enter into a personal relationship with the ultimate creative mind. For religion is not about a purely speculative hypothesis. It is about a life-transforming conscious relationship with a reality of supreme value. In that relationship, finite minds may gain some access to the ultimate mind of all, and their capacities may be enhanced by co-operation with ultimate mind. We might, with caution, even suggest that the divine mind will gain knowledge of new types of experience through its knowledge of finite minds, and will engage in new forms of action through its co-operation with finite mind.

Different religious paths will construe such relationships of knowledge and action in different specific ways. Presumably, only God can enable human minds to know the divine mind, to know and carry out the divine purpose. Diverse religions testify to diverse disclosures of divine presence and portray diverse ways of carrying out the divine will. But in many traditions the central idea is of a reality, or state of supreme goodness, that can transform human lives through their unity or relationship with it and can bring human capacities to an adequate and appropriate fulfilment.

From the perspective of my own spiritual path, it would be said that when God originated finite centres of experience and action, the primordial Spirit renounced the pure bliss of the divine being. For the creation of a cosmos like this will entail much suffering and destruction, God's knowledge of which will enter into the heart of the divine experience. But God obtains in return many new sorts of experiences and values – for instance, values of discovery, learning and friendship – that could only be actualised in the lives of finite agents. God shares both in the joys and in the sufferings of finite subjects, but in the divine experience sufferings are transformed by their integration into a wider experience that sublimates their quality and moderates their intensity.

There can be a sharing of minds, as God comes to know actualised experiences that would not otherwise have existed, and finite subjects are able to access, to various degrees, the wider experience and the inner beauty of the mind of God.

In creating autonomous subjects, God relinquishes total control over all that is. For in permitting real creaturely freedom, God permits things to happen that God would never directly have intended and of which God might well disapprove. But we might think that God will obtain in return the possibility of acting in co-operation with others who have a limited but real power to make their own decisions. God will have new possibilities of creativity, of response to decisions made by others and of concern for the lives of subjects who retain their own unique otherness and individuality. There can be a sharing of actions with others, as God acts both creatively and responsively to empower the thoughts and acts of finite subjects, and those subjects are able to access God's power so as to shape and forward, in their own creative ways, the divine intentions for the future of the world. (This general view is well expressed by the essays edited by mathematical physicist John Polkinghorne in *The Work of Love*, SPCK, London, 2001.)

Other traditions may have different specific ways of spelling out a religious vision of the universe. There is nevertheless a general agreement on what the God of religion is. The scientific God is ultimate mind insofar as it is seen as the intelligent origin of an intelligible, intellectually beautiful and elegant cosmos. The religious God is ultimate mind insofar as human persons believe themselves to have some personal acquaintance with the divine presence, wisdom and perfection and some access to divine creativity, compassion and power. Contemporary science shows the scientific God to be a coherent and plausible possibility. To move on to acceptance of the religious God requires some personal experience that can be reasonably interpreted as access to the presence and power of the divine. You can believe in the scientific God without believing in the religious God. But insofar as they both exist, they are undoubtedly the same God, approached by different paths.

5. THE ULTIMATE GOAL

The scientific question of an ultimate goal of the cosmos is the question of whether there is some set of ultimately worthwhile states in which the evolution of the universe will terminate. A negative answer to this question is suggested to some people by the second law of thermodynamics, by an acceptance of the final inevitable death of the physical universe as we now know it. But some cosmologists have argued that there could be at least a subjective immortality, a form of intelligent life that seemed to be endless, even though it was not. Or it might be held that there could be a flourishing of personal life in the mid-term maturity of the universe, and that the ultimate death of the universe would not matter very much. At least it seems to be widely accepted that ultimately worthwhile states would be states of consciousness, knowledge and happiness, states connected with the existence of conscious and intelligent beings.

Most religious views would agree, but would add two things. First, from a purely physical perspective, such an ultimate goal is going to be achieved by very few, by the small minority of winners in the evolutionary race. Most conscious beings in the universe are going to live with pain, frustration and disappointment, and their best hope will be that some other species in the far future might be truly happy and fulfilled. A religious perspective, which grounds the whole universe in a reality of supreme value, would be more concerned that the possibility of some personal fulfilment could be made available to the whole array of sentient beings, not just to a small fortunate minority. This would require some form of existence beyond this cosmos for most sentient beings.

Second, if there is a God, then the good experiences of finite beings will never be lost, for they will be known and remembered by God for ever. This may seem a good thing, but there is a downside. If many experiences are of pain and frustration, then remembering these for ever is not such a happy prospect after all.

For these reasons, a religious conception of an ultimate goal of the cosmos usually posits that the sufferings of finite creatures will be transformed, not just remembered, by God. Sentient life will not only continue beyond this physical cosmos; it must also be transformed and fulfilled by a deeper and fuller conscious relationship to the Supreme Good, in which physical lives have always been grounded. So for most religions it is important that

God can and will ensure that all finite beings with a sense of a continuing self will be enabled to share in the final flourishing of conscious life that is the divine purpose of the universe.

From a religious point of view, then, the ultimate goal of the cosmos can be seen as the generation of a society of minds, or many societies of minds, sharing experiences and actions, endlessly moving into a future of infinite possibility. The one ultimate mind that is the basis of all other beings may find a new expression of itself in generating and entering into relationships of encouragement, compassion and co-operation with the societies of finite minds that arise from the unbounded potentiality of its own being.

The history of the cosmos can then be seen as the story of a movement of divine self-giving, generating finite persons who, through many struggles, failures, endeavours and explorations, are brought finally towards their fulfilment in God. Ultimate mind relinquishes something of its supreme happiness and power in order to share the sufferings and the consequences of the radically free actions of finite persons, in a world in which suffering and the misuse of freedom are possible and to some extent necessary. But it does so because only in that way can finite persons in such a world be brought, however slowly and painfully, to a state of supremely worthwhile fulfilment and well-being. In the Christian tradition, this is put most succinctly in the teaching of the fourth-century Bishop of Alexandria, Athanasius, that the divine shares in humanity (we might say, more broadly, in the lives of finite beings) in order that finite beings might share in the divine life.

In this sense, as so many mystics and religious teachers have seen, the final meaning of the universe is love. This is not a love that can create beings that never suffer harm. It is a love that sees how suffering is necessarily involved in the creation of sentient beings like us, yet also sees that such beings can finally share in a supreme and distinctive form of happiness. So, in self-sacrificial love, God creates this universe, sharing in the sufferings of creatures within it, and working within the necessary constraints of nature to bring all sentient beings to their final flourishing. Then, in love completed and triumphant, God rejoices in the fellowship of those who have travelled faithfully through the world of time and attained to a life beyond time, to share the divine nature of eternal bliss.

6. SHARING IN ULTIMATE MIND

The speculations of cosmologists are mostly pessimistic about the ultimate future of the cosmos, and in any case it seems unlikely that the flourishing of all sentient beings could be realised within this cosmos. So the cosmos points beyond itself, to a source and to an ultimate goal that lies beyond it. This is a part of the story of the cosmos of which the sciences cannot speak. They speak of what is physical, and the physical once was not and sometime will probably cease to be. Ultimate mind is beyond the physical and will exist long after this physical cosmos has passed away. So, though there may be a purpose for this cosmos that it should end in the realisation of self-aware and self-directive intelligences, such is not the ultimate goal.

In the divine mind all that has ever been is perfectly remembered, though transformed. It seems quite possible that, just as some scientists think that a future intelligence could replicate human persons to live again within the far future cosmos, so the eternal intelligence of God could bring persons to live again in other realms beyond the physical confines of this cosmos. We might expect that a perfect eternal intelligence would be as concerned for every sentient being throughout the history of the cosmos as for any life-form that exists at its end. It may then seem fitting that all such sentient beings that have a sense of their own continuous existence can share in the mind of God, and find there an appropriate sort of fulfilment for what remained incomplete, and a transformation of experience of all that caused pain and suffering, in their cosmic lives.

Thus the ultimate religious goal is a life in God beyond this cosmos. Yet the cosmos remains of vital importance. It is here that distinctive sorts of personal lives are generated, that unique forms of experience and action are actualised and that personal responsibilities for the future are exercised. If the guidance of ultimate mind is real, we might hope that extreme pessimism about the future of the earth will be counteracted by faith in the power of divine providence. We can have faith in the future, because and insofar as there is a supreme creative mind whose purpose is that this earth, and the wider cosmos, should be shaped on the pattern of perfect beauty that lies in its own being.

It is a human responsibility to realise that divine purpose, and so it is a vital obligation to preserve the earth and form it to be a habitat within which understanding, beauty and friendship may flourish. Yet even if we should fail, and though there may always be in this world the tragedy of possibilities unrealised and beauty disfigured, the ultimate goal can still be realised in realms of being beyond this. The way to the goal may be long and hard, because of how far we have set ourselves from the path to it. But, whatever we may have to learn to do and to undo of our cosmic lives, the ultimate goal remains attainable, the patience and wisdom of ultimate mind remain inexhaustible and the attraction of perfect beauty, if it is once seen truly, will be irresistible. That, at least, is the faith that a commitment to ultimate mind inspires.

I hope to have shown that, even in strictly scientific terms, such a religious vision of the cosmos is possible and that the scientific worldview, taken in a positive and not a materialist sense, gives it some plausibility. In the end, however, without personal experience of transcendent mind and without some experiential evidences of the action of such a mind in history (what the religious call 'revelation'), this will remain simply speculation.

To turn speculation into a life-transforming power requires a passionate commitment, made in objective uncertainty (but not in complete ignorance), to the best that we can know. It requires the commitment of personal life, not to some arbitrary and irrational belief, but to the supremely good and beautiful. It requires that we should discern for ourselves in some way and at least to some extent the purifying, illuminating, overwhelming power of Pascal's 'Fire'.

BIBLIOGRAPHY

RELEVANT BOOKS AND ESSAYS BY THE AUTHOR

Books

God, Chance and Necessity. Oxford, Oneworld, 1996.
God, Faith and the New Millennium. Oxford, Oneworld, 1998.
In Defence of the Soul. Oxford, Oneworld, 1998.
Rational Theology and the Creativity of God. Oxford, Blackwell, 1982.
Divine Action. London, Collins, 1990.
Religion and Creation. Oxford, Clarendon Press, 1996.
Religion and Human Nature. Oxford, Clarendon Press, 1998.
God: A Guide for the Perplexed. Oxford, Oneworld, 2002.
The Case for Religion. Oxford, Oneworld, 2004.

Essays

God, Time and the Creation of the Universe (with Chris Isham). London, Royal Society of Arts, 1993.

'God as a Principle of Cosmological Explanation', in *Quantum Cosmology and the Laws of Nature*, ed. Robert Russell, Nancey Murphy and C.J. Isham. Vatican City, Vatican Observatory, 1993.

'Cosmos and Kenosis', in *The Work of Love; Creation as Kenosis*, ed. John Polkinghorne. London, SPCK, 2001.

'The Temporality of God', in *Issues in Contemporary Philosophy of Religion*, ed. Eugene Long. Dordrecht, Kluwer, 2001.

'Believing in Miracles', *Zygon*, September 2002, pp. 741–51.

'Two Forms of Explanation', in *Is Nature Ever Evil?* ed. Willem Drees. London, Routledge, 2003.

'Christianity and Evolution', in *Identity and Change in the Christian Tradition*, ed. Marcel Sarot and Gijsbert van den Brink, Frankfurt, Peter Lang, 1999.

'The World as the Body of God', in *In Whom We Live and Move and Have Our Being*, ed. Philip Clayton and Arthur Peacocke. Grand Rapids, Eerdmans, 2004.

'Cosmology and Religious Ideas about the End of the World', in *The Far-Future Universe*, ed. George Ellis. Philadelphia, Templeton Foundation Press, 2002.

'Theistic Evolution', in *Debating Design*, ed. William Dembski and Michael Ruse. Cambridge, Cambridge University Press, 2004.

'Human Nature and the Soul', in *Science, Consciousness and Ultimate Reality*, ed. David Lorimer, Exeter, Imprint Academic, 2004.

OTHER RELEVANT READING

General introductions to the subject

Russell Stannard (ed.), *God for the Twenty First Century*. London, SPCK, 2000.

Ian Barbour, *Religion and Science*. London, SCM Press, 1998.

Arthur Peacocke, *Paths from Science towards God*. Oxford, Oneworld, 2001.

Chapter 1

On the Genesis accounts:
Keith Ward, *God, Faith and the New Millennium*. Oxford, Oneworld, 1998, chapter 3.

John Weaver, *In the Beginning God*. Oxford, Regent's Park College, 1994.

A short account of the Galileo affair:
Michael Poole, *A Guide to Science and Belief*. Oxford, Lion, 1990.

Chapter 2

On Newton's worldview:
Mordechai Feingold, *The Newtonian Moment*. Oxford, Oxford University Press, 2004.

On compatibilism between physical determinism and freedom:
Richard Swinburne, *Responsibility and Atonement*. Oxford, Clarendon Press, 1989, chapter 3.

On Newton and the laws of nature:
Ian Barbour, *Religion and Science*. London, SCM Press, 1998, pp. 17–23.

Chapter 3

On the general history of the cosmos:
Martin Rees, *Our Cosmic Habitat*. Princeton, Princeton University Press, 2001.

On the relation of God and the universe:
Keith Ward, *Religion and Creation*. Oxford, Clarendon Press, 1996, chapters 7–12.

Chapter 4

On a generally religious view of evolution:
Arthur Peacocke, *Paths from Science towards God*. Oxford, Oneworld, 2001.

On evolution and purpose:
William Dembski and Michael Ruse (eds), *Debating Design*. Cambridge, Cambridge University Press, 2004.

Rebutting atheistic views of evolution:
Keith Ward, *God, Chance and Necessity*. Oxford, Oneworld, 1996.

Chapter 5

On value in evolution:
Holmes Rolston, *Genes, Genesis and God*. Cambridge, Cambridge University Press, 1999.

On suffering and divine purpose in evolution:
John Haught, *God after Darwin*. Boulder, Westview Press, 2000.

On God and evolution:
Arthur Peacocke, *Creation and the World of Science*. Oxford, Oxford University Press, 1979.

Chapter 6

On quantum theory in general:
Paul Davies and J.R. Brown, *The Ghost in the Atom*, Cambridge, Cambridge University Press, 1986.

On the death of materialism:
Paul Davies and John Gribbin, *The Matter Myth*. Harmondsworth, Penguin, 1992.

On God and the new physics:
John Polkinghorne, *Belief in God in an Age of Science*. New Haven, Yale University Press, 1998.

Chapter 7

On issues of physical determinism and an open future:
John Polkinghorne, *Science and Providence*. London, SPCK, 1989.
John Polkinghorne, *Science and Creation*. London, SPCK, 1988.

A readable introduction to process philosophy, which deals with these issues among others:
John Cobb and David Griffin, *Process Theology, An Introductory Exposition*. Philadelphia, Westminster Press, 1976.

On chance and predictability:
David Bartholomew, *God of Chance*. London, SCM Press, 1984.

Chapter 8

On the cosmos as a holistic system within the being of God:
Philip Clayton and Arthur Peacocke (eds), *In Whom We Live and Move and Have Our Being*. Grand Rapids, Eerdmans, 2004.

For general background:
Arthur Peacocke, *Theology for a Scientific Age*. Minneapolis, Fortress, 1993.
Philip Clayton, *God and Contemporary Science*. Edinburgh, Edinburgh University Press, 1997.

Chapter 9

On science and personal experience:
Keith Ward, *The Case for Religion*. Oxford, Oneworld, 2004, part 3.

On whether science tells us everything about reality:
Mikael Stenmark, *Scientism*. Aldershot, Ashgate, 2001.
Mary Midgley, *Science and Poetry*. London, Routledge, 2001.

Chapter 10

On God as ultimate explanation:
Keith Ward, *Religion and Creation*. Oxford, Clarendon Press, 1996, especially chapter 8.
Keith Ward, *Rational Theology and the Creativity of God*. Oxford, Blackwell, 1982.
Richard Swinburne, *Is There a God?*. Oxford, Oxford University Press, 2001.

Chapter 11

On the soul:
Keith Ward, *Religion and Human Nature*. Oxford, Clarendon Press, 1998, chapter 7.
Keith Ward, *Defending the Soul*. Oxford, Oneworld, 1998.
Richard Swinburne, *The Evolution of the Soul*. Oxford, Clarendon Press, 1986.

For general essays on the topic:
David Lorimer (ed.), *Science, Consciousness and Ultimate Reality*. Exeter, Imprint Academic, 2004.

Chapter 12

On the Christian doctrine of 'original sin' placed into an evolutionary context:
Keith Ward, *Religion and Human Nature*. Oxford, Clarendon Press, 1998, chapter 8.

On the evolutionary origins of culture and ethics:
Holmes Rolston, *Genes, Genesis and God*. Cambridge, Cambridge University Press, 1999.
Mark Richardson and Wesley Wildman (eds), *Religion and Science*. London, Routledge, 1996, case study 6.

Chapter 13

On the existence of a Supreme Good:
Iris Murdoch, *The Sovereignty of Good*. London, Routledge, 1970.

On morality:
Keith Ward, *God, Chance and Necessity*. Oxford, Oneworld, 1996, chapter 9.
Keith Ward, *In Defence of the Soul*. Oxford, Oneworld, 1998, chapter 6.

On cosmic ontogenesis:
Pierre Teilhard de Chardin, *The Phenomenon of Man*, trans. Bernard Wall. London, Collins, 1955.

Chapter 14

On divine action:
Keith Ward, *Divine Action*. London, Collins, 1990.
Philip Clayton, *God and Contemporary Science*. Edinburgh, Edinburgh University Press, 1997, part 3.

On God and time:
Keith Ward, *Religion and Creation*. Oxford, Clarendon Press, 1996, chapter 11.
Keith Ward, *God: A Guide for the Perplexed*. Oxford, Oneworld, 2002, chapter 5.

Chapter 15

On faith and knowledge:
John Hick, *Faith and Knowledge*. London, Macmillan, 1966.

On reasonableness in religion:
Keith Ward, *Religion and Revelation*. Oxford, Clarendon Press, 1994.

On the nature of religious belief:
Keith Ward, *The Case for Religion*. Oxford, Oneworld, 2004, especially
part 3.

On miracles:
Richard Swinburne, *Miracles*. London, Macmillan, 1989.
Keith Ward, *Divine Action*. London, Collins, 1990, chapter 10.
Symposium on miracles in *Zygon*, September 2002.

Chapter 16

On endless life in this cosmos:
Martin Rees, *Our Cosmic Habitat*. Princeton, Princeton University
Press, 2001, especially part 2.

On the future universe:
George Ellis (ed.), *The Far-Future Universe*. Radnor, Templeton
Foundation Press, 2002.

On immortal life:
Keith Ward, *Religion and Human Nature*. Oxford, Clarendon Press,
1998, chapters 11–15.

Chapter 17

On the function of religion:
Keith Ward, *The Case for Religion*. Oxford, Oneworld, 2004.

On immortal life, from a Christian point of view:
Keith Ward, *What the Bible Really Teaches*. London, SPCK, 2004,
chapters 4 and 9.

On other religious views of immortality:
Keith Ward, *Religion and Human Nature*. Oxford, Clarendon Press,
1998.

INDEX OF SUBJECTS

Index of subjects

INDEX OF NAMES